LIFE
WASN'T
BORING

Soldiering is a serious, frequently bloody business. That aspect has been documented endlessly. But behind the blood, thunder and de-humanising aspects of conflict are people – people in uniform and people supporting them. All of them have personal feelings and aspirations and they experience the humdrum assortment of daily occurrences that closely match those of their counterparts in civil society. Those aspects of their lives are not widely reported, or appreciated, and it is on those that this book has its focus.

Life Wasn't Boring relates to the life, times, successes, failures and, most importantly, the personal interactions and loves of a professional infantry officer and his family, over more than a third of the century that was his service. Some parts are as serious as can be expected of a military account. Other parts might surprise, entertain and even amuse the reader. Together they hold up a mirror to reveal the human side of being a soldier.

'*I would rather die of passion than of boredom*'
Emilé Zola

Colin Groves

LIFE WASN'T BORING

His life, loves,
laughs and losses
from over three
decades of soldiering

UNIFORM

First published by Uniform
an imprint of Unicorn Publishing Group LLP, 2022
5 Newburgh Street
London W1F 7RG

www.unicornpublishing.org

10 9 8 7 6 5 4 3 2 1

ISBN 978-1-914414-39-8

Typesetting by Vivian Head

Printed in Malta by Gutenberg Press

PREFACE

From the time I had children of my own, I became something of a story teller. When they were young I made up bedtime stories for my sons, Nick and Tim, and many years later, I told more stories to my four grandchildren, Lucas, Ella, George and Louise. Both generations enjoyed them, often asking for more. Additionally, throughout my adult life, I have taken real pleasure in recounting stories, largely drawn from my own experiences, to friends and anyone else who would listen.

For a long time, my family and some close friends have been urging me to record some of my stories and anecdotes that would be lost if I did not write them down. That stimulus was reinforced in November 2019, during a trans-Atlantic crossing on the *Queen Mary 2*, which formed part of the 50th wedding anniversary celebrations for my wife, Sue, and myself. The Cheltenham Literary Festival was the principal on board attraction. I attended a writing workshop, followed by a conversation with Julia Wheeler, an award winning author and former BBC correspondent. She was extremely encouraging and helpful and advised that I attend a residential, creative writing course run by the Arvon organisation. The course was led by Mel McGrath and Jay Griffiths, both accomplished and award winning authors. It was very beneficial and highly motivating.

You have just begun to read the result. Hopefully, the stories in this book will capture your interest. It records aspects of a life that has been wonderfully diverse and fortunate in so many ways.

If you read this, you will probably either be a member of my family, a close friend or military colleague. I would like to thank you for the love, friendship, loyalty and support you have shown me. It has made my life the pleasure it has been and still is. I am deeply in your debt.

Colin Groves
Woodham, Surrey
3 April 2020

CONTENTS

PROLOGUE

I was probably destined to be a professional soldier. Both my parents saw war-time service – my father, Edwin (Ted), a warrant officer in the King's Royal Rifle Corps and my mother, Ruby, a captain and nursing sister in the Queen Alexander's Royal Army Nursing Corps. I was a war baby, born on 26 February 1944, getting my introduction to conflict when only a few weeks old, far too young to have any memory of it.

One of my father's favourite stories was that shortly after I was born he was on leave, visiting his wife and her parents at their home in Kirby Cross. The village is only a couple of miles inland from the Essex coast, which put it under a Luftwaffe flight path for raids on London. He was out for a walk, pushing me in my pram – nothing like one of today's ingenious folding buggies, but a 'big wheeler' that had a tray for shopping under the generous baby carriage. He became aware of aircraft noise and looked up to see a 'doodlebug' high above us, presumably making its way towards the capital. Suddenly there was silence. The doodlebug's engine had stopped. My father saw its nose dip and it began a terrifying dive towards us. The flying bomb exploded in a field nearby. Luckily, no harm was done to us, or the pram. When, years later, my father recounted this story to me, I asked him what he had done to save me.

'Save you?' he said, 'I was under your pram!'

So much for my gallant father's bravery and his overwhelming paternal thought for my wellbeing! Not for one moment did I believe that he took cover under my pram. His knowing smile gave the lie to that claim, but the incident set the tone for my military career – lots of excitement; occasionally more than just a smidgeon of fear; a host of strong, close personal relationships; compelling interest and sometimes real amusement.

CHAPTER ONE
MY FAMILY AND EARLY YEARS

Both my parents left military service at the end of the war. My brother, Rick, was born in 1947 and shortly after his birth our parents bought a laundry business in Llandrindod Wells, in mid-Wales. We stayed there for some seven years and lived in a large, detached house with a huge rear garden and outbuildings that housed the laundry. Immediately opposite was the Shrewsbury to Swansea railway line. *The Railway Children* is a familiar and idyllic story and for Rick and myself, our early childhood was not far removed from it. Of course, Rick was no use at all to me until he was about four, by which time he had learned to run, play games, kick a ball and hold a bat. But once he had achieved all that, we were on the fast track to, more or less, constant fun.

Our parents placed few demands on our movements. We played endlessly. Street games were common, finishing only when it got dark, or a parent called 'bath time'. Before then it was football and cricket, trading marbles, having the 'king' conker, of being in gangs and building secret camps. Much of it involved scratched knees and getting dirty – then the laundry came in very handy. Once we had got our bikes, our radius of action extended for miles, but no one seemed to mind and we regarded our freedom as entirely normal.

By the time I was ten my bike became an essential. I had met my first girlfriend, the dark haired, slender, Margaret Thomas. She was lovely. At an early age we discovered kissing and decided that we liked it. I still think fondly of her. The problem was that she lived in Penybont, a village five miles from Llandrindod. Without the bike it would have been a case of 'love's labour's lost', but my parents were perfectly happy for me to ride over to see her (they knew nothing about the kissing), notwithstanding that the journey was on two 'A' roads, one of which, the A44, was a main arterial route.

We left Llandrindod after our parents' business first merged and then was taken over by a larger concern. The move away from Wales came after I had had only two terms in the 'A' stream at the local co-educational grammar school. I was catapulted into an all boys' school in Liverpool. The environment, and particularly the curriculum, were very different from the ones in Llandrindod and I was at a loss to follow many of the lessons.

My father had become the manager of a laundry business in Formby, in Lancashire. It quickly transpired to be a disastrous move for the family when his new employers failed to keep promises and he resigned after only three or four months to take up a similar position in Atherstone, in Warwickshire. After just one term, I suffered another change of school and I found myself in the second year 'B' stream of the Queen Elizabeth Grammar School, Atherstone.

The school had been established in 1573. Its approach to education was traditional rather than inspiring. It had boys and girls, but numbered only about 350 pupils, small enough for everyone to know everyone. It was a stable establishment, to the point of being unchanging. Most of the teaching staff, who were there on my arrival, were still there when I left and for many years after that.

Other than an imposing roadside façade and assembly hall, most of the rest of the school's accommodation was utilitarian – a flat roofed, two-storey, 1950's build that simply provided space for classrooms, a gymnasium, changing rooms and staff rooms. The school was very compact. The least favourite building was the outdoor, unheated, unfiltered, green watered, permanently cold, swimming pool that could only be accessed through the woodwork classroom. The girls harboured a particular hatred for it, because to have swimming lessons, they had to make their way through classes of leering, woodworking, teenage boys. In contrast, 'The Grove' was splendid, offering a wide expanse of well-maintained playing fields, with football and hockey pitches in the winter and cricket pitches, a grass athletic track and tennis courts in the summer. A grove of very attractive, mature trees gave it it's name. Taken together, the Queen Elizabeth Grammar School was a comfortable, safe place in which to receive an education.

I still had some catching up to do with regard to the curriculum, but after only a short time, I realised that in a 'B' stream I did not have to work too hard to keep my head above water. With only a few exceptions, our teachers did little to promote a real appetite to learn and the school's academic effort was firmly centred on the 'A' stream pupils. I began a long process of coasting that lasted until I left school.

I should not be too critical of the school's offer to its pupils. It was a wide one. Quite elaborate theatrical productions were staged every year. Trips to the theatre and other places of interest were commonplace. There were quite a few after school clubs, including a choir, and the school's sports teams played other schools every Saturday in term time. Occasionally, overseas trips would be organised and I had my first experiences of foreign travel, first to the Dordogne and later to take part in a wonderfully exciting trekking and mountaineering expedition to the Austrian Tyrol.

While for me the academic side of life at school was, at best, middling, its social and sporting aspects were of a much higher order. The close proximity of teenage boys and girls meant many relationships were quickly formed and in most cases, equally quickly discontinued in favour of someone else. Mainly because of the sporting opportunities that were offered, I enjoyed my secondary school years. I was an all-round sportsman and represented the school at football, cricket and cross country running. In athletics I was developing as a better than average sprinter. However, my first sport was football. I represented Warwickshire schools as a goalkeeper and some of our coaching was done at Coventry City, under the auspices of Jimmy Hill, a far-sighted manager and later a TV pundit. I also attracted the attention of scouts from Aston Villa, but after trials, it was apparent that at just over 5ft 8in I was not tall enough to command a regular place in a professional team.

At the beginning of what was to be my last season of living at home, I joined an under-18 side called Young Boys, which played in the Nuneaton Amateur League. Young Boys had only formed a year earlier, but during the course of their first season they failed to win a single match. They had frequently lost by double digit margins and had done very well just to keep going. I reckoned they might appreciate a goalkeeper and that if they wanted me, there would be plenty to keep me busy. The team soon won its first match. I was as busy as I had predicted and my performances were noticed. I was chosen for the 'Pick of the League' side, to compete in a cup competition against sides representing other leagues, drawn from across the Midlands. We got to the final, but lost. In one of the earlier rounds we were drawn against a Birmingham-based league side. The match was played, under floodlights and it had rained heavily. The pitch was awash and early in the match, a low cross flashed across my penalty area. I dived and collected the ball, but aquaplaned for yards through surface water and muddy puddles. When I stood up I looked like the monster from the swamp and had to scrape liquid mud from my eyes and nose and spit it out of my mouth. Billy Harper, our captain, patted me on the back.

'Good save, mate. You know they scattered some bloke's ashes here this afternoon!'

Of course, I could not appreciate it at the time, but that was a foretaste of the black, soldier humour that was to keep me on my toes, maintain morale, mitigate the bad times and make me smile throughout several decades to come. I missed that more than anything else once I left the service.

At home and at school, Rick and I continued to play together. At football I was the more talented, but Rick was good enough to represent the school as a fullback. Throughout our growing up we remained close to the point that when playing football, we could often find each other with passes without really looking. That closeness extended to cricket where he was, by far, the better player – a really good and very fast bowler who could make the ball sizzle in my wicket-keeping gloves. 'Caught Groves C, bowled Groves R' was a frequent entry in many score books.

Our parents both worked groaningly hard running the laundry, so much so that they had little time for family life. They were God-fearing, patriotic and conservative in their values and politics. We were not an affluent family and there were very few luxuries. Holidays were never taken with us all together – one or other of our parents was always working. Both a modest car and ownership of a television came late to us, however, the necessities of life were always provided. Rick and I were deeply loved and well cared for, but we were not a demonstrative family. We simply knew we loved each other and did not feel the need to say or exhibit it. It must have happened, but I cannot recall my parents kissing. While the family may have lacked some overt warmth in its love of one another, the certainty of that love and the support that it brought, were never in doubt. We simply knew it, felt it and were confident in it.

It was against this background that I completed my schooling, or rather I just about squeezed out of it. By the time I was in the Fifth Form and about to sit the General Certificate of Education (GCE) 'O' level examinations, I was doing virtually no work. The results mirrored my woeful lack of focused effort. I managed to pass five subjects (English Language, Maths, Geography, Biology and Art). Maths took me three attempts and Biology two. Not a record to be proud of, but for whatever reason I was allowed to enter the Sixth Form.

It might have been because rarely did a 'B' stream pupil elect to do so. My contemporaries left the school the instant they were allowed to and I never saw the vast majority again.

For my first few years at the school I was aware of Sue Waterhouse, but no more than that. She was in my year group, but in the 'A' stream. However, on our joining the Sixth Form in 1960, we were thrown into closer proximity. We quickly became boyfriend and girlfriend and six decades later we are still together, still boyfriend and girlfriend, which is so much more vital and exciting than just being husband and wife. Our partnership far transcends any other of the successes in my life, but much more of that later.

The two years in Sixth Form were even less demanding than their predecessors. It was agreed that I should take Geography and Art at 'A' level and that was it – no other subjects. I had more free periods than I had lessons. Sue also took 'A' level Geography and we sat at adjacent desks. Atlases were in short supply and during one period she asked if I would like to share one with her. Since then I have always maintained that it was a heavily disguised proposal of marriage that I failed to recognise. In any event, I accepted the offer and the rest turned out to be not so much geography – as history!

I had also made a friend of the sports master, Mr Robin Champion. We shared a birthday and he was one of the relatively few teachers who did motivate and inspire their pupils. I had come to his attention because of my all-round sporting abilities and as a sixth former, he made use of them by allowing me to help with some of the sports and PT lessons that he took. I enjoyed the experience, but it served to take my eye further off my education. Inevitably, my 'A' level results were to be even worse than those obtained at GCE level.

In the last year of Sixth Form some small attention was paid to what profession pupils might follow in adult life. I cannot recall there being a careers' master or mistress. The little advice and guidance that was forthcoming, was very ad hoc. For most pupils the advice only extended to going to university, or becoming a teacher. Both were out of the question for me. I decided that I would try for a commission in the Army, or failing that apply for a position in the Customs and Excise service. I am no longer entirely sure why I came to that decision, but I think I was probably looking for an active, rather than sedentary, working life and one that carried responsibility.

At least two 'A' levels were required to apply for officer training and to gain a place at the Royal Military Academy, Sandhurst (RMAS). However, applicants in their final term at school, were allowed to attend the four-day interview at the Regular Commissions Board (RCB) on the understanding that two 'A' level passes had to be achieved in the examination results that would be published only a few weeks later. In the summer of 1962 that small relaxation of requirement brought me to the RCB, housed in a World War II, hutted, Army camp in Westbury, Wiltshire.

There I was hopelessly out of my comfort zone. The vast majority of the other applicants were public school boys, seemingly confident of themselves and comfortable in their obvious affluence. We were all given overalls and numbered tabards to wear, before being divided into groups of about eight, each group with its own distinctive colour. The four days involved intelligence and general knowledge tests, public speaking, fitness tests, and one-to-one interviews with serving officers and psychologists. They also included individual and team 'command tasks', which tended to involve getting heavy, awkwardly shaped weights across gaps/over walls with the aid of planks that were too short and ropes that were not long enough for the job. When all those aspects were considered together, the four-day interview was an extremely demanding, revealing process and it remains largely unchanged to this day, such is its rigour.

While I was undoubtedly unsettled by my surroundings and fellow applicants, I did have some advantages. I scored very well in the intelligence tests and did even better at general knowledge. My trump card at RCB was my athleticism and fitness. I was able to scale whatever was put in my way, run faster and jump longer and higher than most of my fellow applicants and I could see that that made a favourable impression. We all left the ordeal with the knowledge that the outcome of our efforts would be sent by post within a couple of days. If the envelope was a fat one, we had passed. A thin one contained 'thanks, but no thanks'. When mine arrived – it was thin!

That was a bitter disappointment, compounded a few weeks later by my failure to pass even one of the two subjects that I had taken at 'A' level. To compound my misery, I even managed to fail my driving test. That dire catalogue of repeated failure acted as a forceful wake-up call. A re-think and re-doubling of effort was required. I looked hard at myself. Despite the dreadful academic record, I knew I had a quick brain when I chose to use it. I could readily accept responsibility (I was amongst the first in my year group to have been made a school prefect) and already I could recognise a nascent ability to organise people.

I considered the Customs and Excise again, but it appeared a poor second best to the Army. However, there was a last straw to grasp and I determined to grab it. The Army allowed two attempts to pass the RCB. That much was made clear in the letter contained in the 'thin' envelope, but it also provided the advice that before re-applying, most people would benefit from finding some constructive work for a couple of years in order to gain experience and maturity. I was in no mood to wait and six weeks after failing I was back at Westbury for my second attempt.

This time Sandhurst and a regular commission could not be my goals. My failure to secure two 'A' levels precluded that, but my five GCEs brought me the chance to apply for a short service commission that would last for three years, after six months of intensive training at Mons Officer Cadet School, Aldershot. Despite that major change, the interview procedure the second time around was exactly the same as the one I had undergone a few weeks earlier.

I was much better orientated for the second attempt, but what I did not expect was that most of the applicants were university graduates, several years older and markedly more mature than those I had met during my first visit to the RCB. I simply threw myself into the interview procedure. Again fitness, intelligence and general knowledge were strengths and I knew much better now how to conduct myself during the very testing 'command tasks'. When in command I was much more assertive than I had previously been and when in a supporting role, I tried to judge when to take direction as given and then appear at the forefront of those making a big effort. I just gave it all I had got. Nobody leaving RCB has any real idea of how successful they have been and so the wait for 'the letter' to arrive seemed a long one.

The following Saturday morning I was still in bed when I heard the letter box rattle. I jumped out of bed, but my mother was there well before me.

'It's a fat one!' she called up to me.

I had passed. I was going to join the Army and, if all went well, become an officer.

Things were moving fast. The 'fat' envelope not only had the good news of my success at RCB, it also contained instructions to join Mons Officer

Cadet School only a few weeks later, plus a host of things to do to get ready for training. It even had a list of civilian clothes that potential officers would need that included a trilby hat! I immediately set about following the fitness regime that the joining instructions advocated that centred on building stamina through five to ten-mile runs repeated daily and exercises to develop upper body strength. I also bought a decent sports jacket, blazer, a second suit and the trilby hat; and opened a bank account with a very generous £50 that my parents donated. I was all set, or so I thought.

CHAPTER TWO

OFFICER TRAINING

PART ONE – A FIRST ATTEMPT

MONS OFFICER CADET School was in Aldershot, in Hampshire, a relatively small town of about 25,000–30,000 people, some 40 miles south west of London. Its claim to fame was, and is, that it is 'the home of the British Army'. In the 1960s, the garrison there was probably the biggest in the UK. The military estate was enormous and contained the units of the 16th Parachute Brigade as well as a host of training establishments. Outside the urban areas of Aldershot, large parts of the countryside are natural heathland, ideal for military training areas and live firing ranges. 'Aldershot' and 'Army' have been synonymous for decades, if not centuries. I would be entering a 'khaki world'.

The few weeks of preparation at home passed quickly. Quite early in the morning of 5 November 1962, I caught a train in sufficient time to arrive in Aldershot well before the reporting deadline. I said my goodbyes to my family. I knew my parents were proud that I had got even this far and they were as hopeful as I was that I would do well.

During the journey I became increasingly aware that I was beginning the transition to adult life. Despite an outward bravado, a lack of confidence meant that it was only after I arrived in Aldershot that I dared put on my trilby. Other than a school cap, it was the first time I had worn a hat and I felt really self-conscious.

There were no taxis immediately available at the station. Knowing that I had a lot of time in hand, I decided to walk. That was a serious mistake. Despite its Aldershot address, Mons Barracks turned out to be about three miles from the station and, in those days, suitcases lacked the wheels that they have today. By the time I got to Mons I was extremely hot and a little dishevelled. It was not at all the sort of impression that I had planned to give.

A host of red-sashed, extremely loud, senior non-commissioned officers (SNCOs) lay in wait for the new arrivals. We were quickly shown to our accommodation. That turned out to be a sprawling complex of co-joined wooden huts called 'spiders'. In the centre were the washing facilities and toilets, along with drying rooms. Radiating from the centre were long, dormitory (barrack) rooms that each accommodated about twelve people. We were shown to our bed spaces, which were to be our home for the next twenty weeks.

A 'bed space' was some six feet wide and about as long. It contained a metal-framed bed, a single wardrobe, a chest of drawers, a bedside locker, a small square of matting and a chair. Each item had its own particular place and there was no decoration – no room in that room for any individuality. One of the SNCOs, a whippet thin, acutely alert man, introduced himself as Sergeant Mason of The Worcestershire Regiment, our platoon sergeant. He told us, in no uncertain terms, that he was to be our principal guide and mentor during our training. Our first duty was to 'fall in' as a platoon, at a particular time in the afternoon when everyone should have reported for training. Although most of us had only a vague idea of what 'falling in' entailed, we soon found out.

First came a lot of shouting and Anglo Saxon demands to 'move your fucking self in double time, sir'. It took a while to adjust to the level of swearing, which was part of everyday parlance in the Army. For many, many soldiers, most nouns rarely stood alone and 'fucking' was an almost mandatory adjective. And nothing was ever broken – it was always 'totally fucked'. Surrounded by such language, it was a sad truth that bit by bit, it entered one's own vocabulary, but as soldiers work in a violent environment, it is unsurprising that the language they use is equally forceful.

Once outside our accommodation we were required to form up in three ranks by platoons (a platoon comprises about thirty soldiers). It transpired that each platoon had its own designated barrack rooms, so we were to train, eat and sleep in that formation throughout our training. My platoon was 5 Platoon. There were three platoons on parade and together they formed Kohima Company.

The platoon sergeants kept bawling instructions until they appeared grudgingly satisfied with the result. Then enter stage left the enormously impressive figure of Warrant Officer Class 2, Company Sergeant Major (CSM) Jennings, The Queen's Royal Surrey Regiment. He was immaculately turned out and radiated huge confidence and authority. He had a voice like thunder and wished us a 'Good afternoon' before promptly informing us that although officer cadets were to be called 'sir' by non-commissioned training staff, we were to call all warrant officers 'sir', the difference being that we were the ones that meant it! No one disagreed. We then 'shunned' and stood at ease for a considerable number of times, as a foretaste of what real drill would be like when we encountered it the next morning. Finally, we were marched off to collect our bedding, cleaning materials and the first instalment of our uniforms.

Once we had struggled back to our barrack rooms laden with blankets, sheets, pillows, pillow cases, mattress cover and a linen bedspread, our platoon sergeant introduced us to the 'bed block'. We learned that beds were only to be made down when we slept in them. At all other times the items of bedding had to be made into a bed block that had the sheets folded with minute precision and placed between two, equally precisely folded blankets. The whole thing was then tightly wrapped in a third blanket to look like some multi-layered sandwich (i.e. the bed block). That was then placed on a fourth blanket, which was stretched over the bed, so tightly that a coin would (had to) bounce when dropped on it. The whole arrangement was topped off by two pillows with their cases folded to match exactly the dimensions of the sheets and blankets beneath. It was a work of art, not easily mastered, and we had to have that done by the following morning.

Next we were instructed on cleaning the barrack room and our share of cleaning the central facilities. The instructions were as precise as those for the bed blocks and we all thought that the standards of cleanliness were set ridiculously high. Any porcelain must gleam; brass shine to the point of reflection; and the linoleum floor appear as a brown ice rink. It was going to be a long night. Only then were we marched to the large, echoing, austere and sparsely furnished cookhouse for our first Army meal – Formica topped tables and plastic stackable chairs were very much in vogue. The menu consisted of simple British dishes, but the food was well cooked, nutritious and there was a wide choice. There was a lot of it and we were to need it in the coming weeks.

Back in the barrack rooms we practiced making bed blocks, with minimal success for most, and then turned our attention to our new uniform, just denim overalls for the first few days. The ironing was easy for me, with laundry experience behind me. 'Blancoing' was a different matter. We had been issued a web belt and gaiters that required blancoing. Blanco was a sort of paste that had to have water added to it, before it was applied to belts and gaiters with a small brush. When it was dry, the colour had to be absolutely even, but it had an unbelievably delicate surface – the merest touch, normally when the belt was being assembled, left a mark and the process had to be begun again for the item to pass inspection. Only when all that was completed satisfactorily could we attend to cleaning the accommodation. Hours later, at about midnight, we thought we had made a fair attempt and turned in.

We were awoken only a short time later by, perhaps, eight SNCOs, in their red sashes, running through our barrack rooms and roaring at us to get outside in whatever clothes we had time to put on. Again we were formed into our platoons and were doubled away to be given a night-time tour of the camp. The tour went on for quite some time before we were marched into an enclosed compound. The shouting died away and gates closed behind us. As soon as that happened, many of the windows of the huts in the compound flew open and women started screaming at us to get out of their accommodation area. These were women soldiers of the Women's Royal Army Corps (WRAC) and they were hugely enjoying the confused embarrassment of semi-clad, brand new, officer cadets. We had been had. We extricated ourselves as quickly as possible and ran back to our barrack rooms. They had been trashed! All that cleaning now counted for nothing. The apparent SNCOs were members of the senior intake of officer cadets and this was our initiation.

My particular corporate cleaning task was to polish three corridor windows and the brass handles that went with them. A smoke grenade had been let off underneath them and they were covered in sticky, green residue. More hours of washing down and polishing followed and we saw no more of our beds that night. I was woefully inexperienced and defenceless and that first dawn was one of the lowest points of my service. I was close to tears and felt a very long way from home. I had to fight down the urge to leave. Things had got to get better.

Predictably, the morning room inspection did not go well, but the senior intake had overstepped the mark by some distance and, I believe, the ring leaders were very severely disciplined. Virtually all of the new intake were dispirited to some extent and all of us were struggling to contend with our new environment, but we were not given time to brood. After breakfast and the unsatisfactory room inspection, it was time for our first 'muster parade'.

Our personal turnout was inspected and then we were introduced to the first elements of drill. We were hopeless, or so we were told ... and told ... and told again. Next came a second visit to the stores to collect the balance of our uniforms and equipment. Each cadet staggered back to his barrack room under a mountainous pile. It was here that we began to appreciate just how much we were to rely on our platoon sergeant throughout our training.

Sergeant Mason would be with us from the time we got out of bed in the morning until late into the evening directing, cajoling and advising. He taught us the more obvious military skills, but he also taught those that needed it, to master the art of ironing. Predictably, it was clothing to begin with, but shortly after, believe it or not, it was our issue boots. The boots were made of pimply leather that needed to be ironed smooth if they were ever to shine – no, gleam – and with a deep, reflective quality. He explained how the various pouches, belt and yoke fitted together to make our combat webbing. He showed us how to clean our kit and, later, how to maintain weapons and present them all to the exemplary standard that was demanded. He brooked no slacking, but encouraged any effort deserving of praise. Right from the beginning I realised the worth of NCOs, particularly SNCOs. As a generalisation, NCOs are the Army's foremen and women. They ensure that daily routines run efficiently, while officers are much more involved with setting the overall direction of activity. SNCOs are remarkable people; strong, knowledgeable characters, skilled at their trade and invariably steadfast and loyal to a fault. They are the backbone of a decent army and any sensible officer recognises their strengths and values their worth accordingly.

We were about to embark on five weeks basic training that all newly joined recruits undergo, but, understandably, so much more was expected of potential officers. During that five-week period we were all confined to barracks. There were to be no distractions from the business in hand.

Basic training is very much the province of NCOs and officers are best kept at a distance. Later, when I had joined my battalion, very occasionally a NCO would quietly say 'N-O-B sir'. It stands for 'Not Officers' Business'. Invariably, in those circumstances, NCOs know what has to be done and are best left to get on with it. An early lesson for any officer was to recognise when your presence was not needed or appreciated. Such absence, judiciously applied, demonstrates trust and builds respect on both sides.

The pace of the next five weeks was breathtaking, frequently literally. The programme progressed frenetically and focused on drill, building fitness and stamina, weapon training and the first elements of field craft and radio procedure. All the while we were getting used to the endless rounds of cleaning and polishing, learning that 'OK' was nowhere near good enough and that we were to strive for perfection in everything we did. We learned to operate as a team and drill has an important part to play in that. Responding instantly, as a formed body, to a command and doing so with perfect synchronicity is a remarkably effective team building process. It does not produce unthinking automatons, as some detractors would have. Rather it produces a very efficient machine, with each component knowing how and being able to play its part. It is the foundation on which much military training is based.

After the first few days, I had settled sufficiently to take stock of my surroundings. There was instant camaraderie amongst the new intake, possibly because we perceived our situation to be a common adversity. I quickly recognised that all of my new found colleagues were people of substance. The Regular Commissions Board had already proved to be an effective winnowing process. However, I also recognised that the British component of the intake closely reflected that of the second interview at RCB. Virtually all of my fellow officer cadets were post graduates, or had had some other experience of adult life. A few were older still and had been outstanding NCOs, recommended for a commission by their units. They too had passed RCB. But there were only three or four other teenagers, like me, on the course. We young ones lacked the maturity and life experience of our older counterparts and, as at RCB, both those qualities would stand them in good stead as the course progressed.

There was a large Commonwealth contingent amongst the officer cadets, mainly from African and south east Asian nations. Together they formed over twenty per cent of the intake. All of them were serving soldiers, especially selected by their armies for officer training in Britain. While their military skills were an advantage, they were to suffer physically.

This was the beginning of the winter of 1962–63; one of the worst winters on record. By the time that we had completed our basic training the temperature had plummeted and the first snows had arrived. The Commonwealth cadets became increasingly anxious and were permitted additional blankets if they needed them – the accommodation huts were barely insulated and the primitive central heating system was woefully inefficient. A huge Nigerian cadet slept in the bed space next to mine. He suffered incredibly and occasionally cried from the cold. One day he came back from the stores with his arms full of blankets. None of us could see what he could do with them all. His solution was to get into his wardrobe at night and have the blankets packed around him. Heaven only knows how he slept, but he maintained that extraordinary ritual for several weeks until the weather was warmer.

Extra blankets were of little use to the Commonwealth cadets as basic training was completed and the course moved on to tactical training, involving more and more field work. Our officers became increasingly in evidence as tactics was their territory. The only good news was that the bed blocks had only to be made once a week, to be included as part of a minutely detailed inspection of our barrack rooms.

In parallel with all this, there was a certain amount of administrative activity to be completed. Early on, all officer cadets were called to their platoon's office to take the Oath of Allegiance and to 'sign on' for the three years that a short service commission would last. Most officer cadets transferred their bank accounts to one of the two army agents – either Glyn Mills or Coutts – banks that specialised in dealing with the Army and officers in particular. During my service, I was to need Glyn Mills' ready understanding, because for much of it we were dreadfully underpaid. However, the most important task was for officer cadets to find a regiment, or corps, to join.

We were instructed to list three regiments, or corps, in order of preference and then representatives of those regiments/corps would conduct acceptance interviews over the next few weeks. I was fairly certain that I wanted to become an infantryman. Like many who knew only a very little about the Army, I thought about putting the Parachute Regiment as my first choice, but my Platoon Commander advised that their representatives would probably conclude that I was too young (immature) to be considered seriously. I also lacked any connection with that Regiment. As my father had served in the King's Royal Rifle Corps (which now formed part of the Royal Green Jackets), they seemed like a good bet and I made them my first choice. I had been born in Essex, so my second choice was the East Anglian Regiment, and having lived in Wales for several years, my list was completed with the South Wales Borderers.

The interview, with senior representatives of the Green Jackets, did not go well. They displayed only passing interest in my father's service and told me that they thought a former warrant officer's son might find life 'uncomfortable' after joining the Regiment. I was taken aback at the unabashed snobbery and although the course left little time to dwell on it, the snub did not go away and remained smouldering within me. On the other hand, the colonel who interviewed me for the East Anglian Regiment could not have been more encouraging. Immediately he latched on to my county connection. He delved a bit and was even more pleased to learn that my maternal grandfather had served in the Suffolk Regiment in World War I. The fact that my grandfather had been a private soldier and not an officer, did not matter in the least. The East Anglians were, and still are, very family orientated and I was offered a place. I also obtained a place with the South Wales Borderers, but was pleased to accept the earlier offer from the East Anglians. I could not have made a better choice.

News travelled fast. Within days of me accepting a place in the Regiment, I got a letter from the regimental tailor congratulating me and advising me of when their representative would next be at Mons. Apparently they were looking forward to making my more formal uniforms. The Army supplies its officers with their combat and working dress and pays a one-off grant for their other uniforms, i.e. those used for formal occasions like parades and mess dinners (No 1 Dress (blues), No 2 Dress (khaki) and Mess Dress (scarlet jacket with yellow lapel facings for the East Anglians). Once the initial purchases are made, it falls to individual officers to maintain those uniforms throughout their service and to buy new ones if required. It took four fittings before the tailor was satisfied with the results. It did not help that with all the physical exercise involved in training, I was changing shape almost week by week as I gained strength and bulk.

Training was relentless. We soon experienced our first, extended field exercise, 'Exercise Marathon Chase', held locally on Hankley Common, a training area that we would get to know all too well in coming weeks. It entailed digging slit trenches and shelter (sleeping) bays, with the small spade, or pick, that fitted onto our webbing. Once that was done, patrols had to be mounted, enemy positions identified and their attacks repulsed. When we had successfully done all that, we filled in our trenches, marched for miles to another position and did it all over again. The 'enemy' was provided by The Black Watch, who provided a resident demonstration platoon to assist with training. They had a fearsome reputation, amplified by alarming stories told by our training staff – no one wanted to be captured by the Scots. The exercise lasted for about four days. There was virtually no sleep and the weather was bitterly cold, although there was no snow. When the exercise concluded we marched more miles back to barracks, where the first task was to clean our weapons, then equipment and only then ourselves. All that done, we were left to recover. I intended to go for the evening meal, but took the opportunity to lie down on my bed for a few minutes. I woke up fourteen hours later and was not the only one to do so!

Some six or seven weeks had now passed. We were basically trained soldiers. We could handle our weapons efficiently and knew how to look after ourselves, our kit and accommodation. The pace was stepped up and more and more the emphasis turned to military tactics and procedures, and the ability to command. Lecture followed lecture and many were reinforced by practical demonstrations by the soldiers of The Black Watch. By now snow had arrived with a vengeance and getting out of barracks to the training areas was difficult as many roads were repeatedly closed. On the training areas the snow lay largely undisturbed and very deep. The Black Watch often struggled waist deep to demonstrate formations and procedures for crossing country, patrolling, defending, withdrawing and attacking. While the weather slowed our progress, it was never allowed to stop it. It was hard going.

There was great concern amongst the warrant officers and SNCOs because the senior intake were about to be commissioned, in what would be something like our week ten of training. It called for an elaborate parade and the drill of the junior intake had to reach the standard required for such an auspicious occasion. The deep snow threatened the parade and slowed progress with drill, but there were large sheds where drill could still continue, albeit in a limited form. In the event, the commissioning parade was held in the largest of the sheds, which was a great disappointment for the cadets of the senior intake and their families. I do not think their disappointment lasted for long. The cadets were delighted to have finished their training at Mons and to have a second lieutenant's 'pip' on their shoulders.

We were now the senior intake. We were kinder than our predecessors and the new 'juniors' probably got away lightly. Most of us had more serious matters on our minds because the intensity of training reached a new level.

Now that we had gained an understanding of the procedures that governed the various phases of warfare (e.g. defence and attack), the focus was firmly on the cadets' potential to become competent leaders. To begin with we were required to command about eight men – a section. Exercises might involve a patrol, an attack, a defence or a withdrawal. Nominated cadets had to give the necessary detailed orders, rehearse their section and then command their contemporaries during the exercise. It placed all but the most competent cadets under a lot of pressure. Many mistakes were made and the criticisms were frequently harsh, albeit accurate and justified. The goal of a commission often seemed a very long way away.

On a personal level, I was good in the field. When it was their turn to command, I was regularly chosen by my fellow cadets for the more difficult tasks, like the lead scout, or route finder, but my leadership abilities were much less well developed. I was the youngest in my platoon and I found it difficult to impose myself on those who were more worldly wise. That tentativeness did not impress the training staff and I was given a warning that I would have to improve, or I might suffer relegation to the junior intake. However, I was assured that I was not being considered for sacking.

Sacking was brutal and came once a warning had been given and a cadet's consequent performance remained so poor that a commission was not a possibility. At morning muster parade a name would be called and a cadet would be told to remain behind while the rest of us were marched off to begin the day's training. That would be the last we saw of the poor unfortunate. By the time of the mid-morning break, he would have left Mons and the Army. His bed space would have been removed and there would be no trace of his ever having attended the course.

About two thirds through (Weeks 13 or 14) the course entered its most intense phase and a military knowledge examination loomed. It had to be passed. A second attempt was allowed, but another failure meant immediate relegation. In the event, I was delighted not only to pass, but to do so in the best half dozen of the intake. Given that so many of my fellow officer cadets had degrees and I had not a single 'A' level, I felt I was not such a dunderhead after all. It was quite a boost.

Field exercises become more and more frequent, lengthy and demanding as the tactical scenarios were altered for cadets to practise command at platoon and finally, company level (about one hundred men). They were staged on training areas throughout England and Wales. All the time the weather remained a challenge – the Brecon Beacons in March were not for the faint-hearted. Most of us got used to contending with being very wet and very cold, but many of the Commonwealth cadets were unable to make the huge physical adjustment required of them and their removal from exercises to recuperate became a commonplace event.

My command abilities had improved, but I was aware that I was still a borderline case and relegation hung over me. My turn to command at platoon level was probably to be decisive. When it came, it involved a tricky night withdrawal and a move, over several miles, to occupy a defensive position. The night was very black, but I had made a sound plan and had given a reasonable set of orders. Things went quite well and after the debriefing by the training staff, I was quite encouraged, if not wholly convinced that I had escaped the drop. However, that was not to be. I was called to the platoon office and was told that I was to be relegated to the junior intake. The course had reached Week 19 of its 20 weeks – so near and yet so far. It was a crushing blow for a nineteen year old to sustain.

Relegations were not made lightly. After the interview with my platoon commander, I was required to appear before the School's commandant, a brigadier. Immaculate in my best uniform, I paraded outside his office. I was astonished to find that all the nineteen year olds on the course were among the six or seven paraded there. One by one we were marched in. It was not that I had done particularly badly. Rather that the training staff thought that going through the last nine weeks of the course for a second time would significantly increase my confidence and that would be vital to such a young officer. It turned out that the same logic was sensibly applied to all the teenage cadets, but it was a difficult pill for any of us swallow.

Throughout my time at Mons, I was the goalkeeper for the School's football team. I was the only officer cadet on the team that was otherwise formed by soldiers of the permanent staff. Rugby is very much the officers' game and to have an officer able to play football well is still unusual. The brigadier's final remark to me was that if I could show just some of the confidence in command that I showed on the football pitch, I would have no trouble in passing the course. I do not know why, but I smiled and the brigadier, quite pointedly, asked what I found so funny. I said that it had taken me nineteen years to develop my goalkeeping abilities and I hoped it would not take so long for me to complete the course. Thankfully, the brigadier laughed and congratulated me on retaining a sense of humour. I think he saw a side of my character that had previously eluded him.

After that interview, all that was left for me to do was to clear my bed space; pack my kit and hand it into the stores for safekeeping; say goodbye and wish my hitherto contemporaries every success on their commissioning (including two, John Hart and Tony Winton, who were to join my own Regiment); and depart for home and a week's leave. I would be spared the pain of watching my friends march off to a commission from the ranks of the junior intake. Nevertheless, I felt pretty wounded.

PART TWO: THE RE-RUN

The week at home with my family helped enormously. They were as understanding and encouraging as I knew they would be. Their support was heavily reinforced by Sue. Since leaving school she had found a job in the local library, simply by walking in and asking for one – how times have changed! She was immediately in her element and after only a short while, she left both the library and home to study at Loughborough and qualify as a librarian.

I returned to Aldershot refreshed, but not really looking forward to joining my new platoon in Salerno Company, with the stigma of relegation attached to me. However, serendipity took a hand to ease my way.

Just inside the entrance to every barrack room was a small, partitioned room intended to afford some privacy to junior non-commissioned officers (corporals or lance corporals) who would be in charge of a barrack room's occupants in a normal unit. When I arrived in 2 Platoon, the occupant of the small room, in the barrack room to which I had been allocated, had exercised his right to leave the Army having decided that it was not for him. As all the other bed spaces in the barrack room were occupied, I was given the small room. With it came responsibility for overseeing cleaning. It was probably a smart move by my new platoon officer and sergeant because it gave me some modest authority, on an ongoing, daily basis and that small step helped me to get used to being 'in charge'.

Once training recommenced, I immediately recognised what I had experienced only a couple of months previously. It made things so much easier. It was now a matter of polishing performance, rather than assimilating knowledge and putting it into instant practice. As the weeks progressed I grew in confidence and that increased assurance was reflected in improved performance.

The weather also improved with the arrival of the spring and summer months. The rigours of winter training were a thing of the past and that too relieved a lot of pressure. Football gave way to athletics. I represented the School as a sprinter and was selected as a reserve for the Army team. That gave me a certain kudos and added to my growing self-assurance.

However, there was a complicating factor for everyone at Mons. At the end of the course, the reviewing officer at the commissioning parade was to be none other than The Queen. Preparations were already underway by the time I rejoined after my leave. Paint was being liberally applied to any surface that Her Majesty might pass. The parade square was re-tarmaced. Grass was not so much cut, as manicured and we drilled and drilled and drilled. Mons Officer Cadet School did not have the impressive backdrops of Horse Guards or Sandhurst, but our sergeants were determined that The Queen would see no diminution to the standard of drill that she normally witnessed.

The final exercise came and went without a hiccough. The list of officer cadets to be commissioned was published and on it was 'Officer Cadet C. Groves – 3rd East Anglian Regiment (16th/44th Foot)'. I had made it. Despite it being a second attempt, I would still be only a little over nineteen years old on becoming an officer.

The day of my commissioning was 26 July 1963. All cadets were allocated two guest tickets for the parade. My parents were the obvious choice, but, to my bitter disappointment, my father decided that he could not leave his work. Sue took up the second ticket. It was an idyllic summer day. The commissioning company marched on to the parade square to take up its position in front of the junior intake and directly in front of the dais where The Queen would review the parade and take the salute. Her car made a stately entrance and she got out accompanied by the then Secretary of State for War, the Right – but not so Honourable – John Profumo. His affair with Christine Keeler had yet to break, but in retrospect, he added colour to, what was for us, a momentous occasion.

After the parade, there was a miniature Royal garden party for The Queen and all the guests, held in a large marquee. I could see my mother was very proud and Sue was pleased for me, although obviously out of her comfort zone. It was the first time that she had come face to face with the Army. As both my mother and Sue had to return home after the garden party, I had had to find a partner for the commissioning ball in the evening.

In the event that was easy. I accompanied the sister of a fellow officer cadet, whose parents had been kind enough to invite me to stay at their home for several weekends during my second stint of training. The ball was held in the marquee and was, by far, the most glamorous event that I had ever attended. All the officer cadets wore their new mess uniforms for the first time and at midnight we were commissioned. The noise was deafening and celebrations continued until dawn. Having said goodbye to my partner, who I was never to see again, I returned to my barrack room. No one went to bed. We carried on drinking something approximating champagne until it ran out. Then somebody had a 'good idea'.

One of our fellow cadets had drunk far too much. He was on his bed, snoring loudly and he could not be woken. One of the soldier truck drivers was persuaded to bring his truck to the accommodation. We loaded the unfortunate sleeper, still in his bed, onto the back of the truck along with his wardrobe, chest of drawers, chair and square of matting. All of it was then transported to the roundabout at the top of Queen's Avenue – an imposing boulevard that runs straight as a die for perhaps two miles through Aldershot's Army estate. The sleeper was unloaded onto the roundabout and his bed space was recreated, just as it would have been at Mons. We left him there. He would probably have regained consciousness to the sound of honking car horns. He was brought back to Mons, an hour or so later, in a civilian police car. The police were very understanding provided that the bed space was removed as quickly and as efficiently as it had been established. Aldershot police are used to dealing with soldiers!

After a really good and much needed breakfast, all that was left to do was pack our personal kit; assemble that which had to be returned to stores; say a heartfelt and very sincere goodbye to the training staff; wish each other the best of luck and head off for leave and whatever lay in front of us.

My last memory of Mons is of shaking hands with Sergeant Jackson,

Coldstream Guards. He was my platoon sergeant. I thanked him warmly for his advice and guidance. He congratulated me, then took one pace back, saluted and made a great play of asking my permission to carry on. I returned his salute and said, 'Carry on please'. He turned to his right and marched off. It was my first salute. A big moment – my commission had been recognised.

CHAPTER THREE
BEGINNING IN BALLYKINLER

MY COMMISSIONED LIFE started in Ballykinler, in Northern Ireland. On the military front, the IRA's largely forgotten Border Campaign had recently concluded. It had begun in 1956, but lost momentum and ground to a halt in February 1962. Security was firmly in the hands of the Royal Ulster Constabulary and so the Army that I joined in Northern Ireland was very much in reserve. However, despite the sense of normality that prevailed almost everywhere, the threat from the IRA had not gone away entirely. Whenever a formed body of troops left barracks, it was armed and carried ammunition, so that it was capable of immediate response. In the event, Northern Ireland remained peaceful and old fashioned garrison soldiering was the order of the day.

I only spent a few days at home before packing my bags and an officer's trunk – a throwback to the campaign days of yesteryear; the sort of thing first class passengers on a Cunard liner would have had at the turn of the twentieth century to transport their clothes and it was an ideal container for my new uniforms. I headed for Heysham, in Lancashire, for the overnight ferry crossing to Belfast. I was to join the 1st Battalion, 3rd East Anglian Regiment (16th/44th Foot), nicknamed 'The Pompadours', stationed in Ballykinler in County Down, some 30 miles south of Belfast.

The battalion's designation might well be unfathomable to a lay person. The county regiments, with which some may be familiar (e.g. the Middlesex Regiment, or the Suffolk Regiment), were being amalgamated into larger regiments in the late 1950s and early 1960s and the 1st, 2nd and 3rd East Anglian Regiments were in the forefront of that process. However, each of the county regiments had well over 250 years of distinctly individual history and it proved extremely difficult to let go of those separate identities. Infantry regiments were formerly Regiments of Foot and before assuming county titles, they had been known by a number. In essence, the 16th Regiment recruited from Bedfordshire and Hertfordshire. The 44th Regiment had its base in Essex. The 3rd East Anglian Regiment still drew the vast majority of its soldiers from those three counties, thus a local, 'family' feel was maintained, albeit in a clumsy title.

The full title was not often used. We were 'Pompadours' and were known as such. The nickname came about because a forebearer regiment adopted purple facings (lapels) to its red coats. At the time, Madame de Pompadour was in vogue and she very much favoured purple, thus the regiment that sported this distinctive colour on its uniforms was given her name.

On boarding the ferry I went to the purser's desk to ask for my cabin key.

'Yes, lad,' said the purser, 'what's your name?'

'Second Lieutenant Groves,' I replied.

'Really?' he said, 'I thought you were still at school.'

Mental note – shave twice a day to encourage a manly shadow and work on gravitas!

Next morning I stood on a Belfast dockside by my pile of baggage. An Army Land Rover was parked a few metres away and a sergeant, in uniform, got out and came over to me.

'Mr Groves?' he asked (subalterns – 2nd lieutenants and lieutenants – are always referred to as 'Mister' rather than by their rank).

'Yes,' I said.

'I'm Sergeant Hutchinson, your platoon sergeant – pleased to meet you, sir.'

Sergeant Hutchinson was a small, slight man. He had a quiet voice and disposition – very different from the training sergeants at Mons.

'We'll get this kit loaded and be off to camp.'

I bent to pick up my bags.

'Don't worry about that, sir,' he said, 'I'll get a recruit to do that for you.'

With that he beckoned to a young soldier standing by the Land Rover, who had also arrived on the overnight ferry, 'Come and move this officer's kit.' The soldier doubled over to us.

'What's your name?' I asked. 'Private Taylor, sir,' he said.

We were not to know it then, but Graham Taylor was to serve with me, off and on, for the next three decades. We played football together in the Battalion's team for years. He was an outstanding soldier and was eventually commissioned, to retire as a captain. In the first five minutes since arriving in Belfast, I had met two people who would become trusted friends and have an enormous influence on me and my career. What a great start!

On arrival in Ballykinler I discovered that it was no more than a hamlet, a remote place, positioned immediately inland of sand dunes on the northern side of Dundrum Bay. The civilian community, such as it was, was dominated by two Army camps. One was brick built and served as the home of the regular infantry battalion. The other was a sprawling conglomeration of wooden huts, a training camp that immediately adjoined a series of ranges. The sand dunes, beaches and the majestic Mourne Mountains, on the opposite side of the bay, made a stunningly beautiful landscape, but that was the extent of the place.

I was given no time to admire the countryside. The adjutant (a captain, who ran the commanding officer's office and who was responsible for discipline within the unit) required me to parade in all my uniforms to ensure that they were of the right standard. I was to need the more formal uniforms virtually immediately.

A mess dinner was held on my second night with the battalion. Two days after that there was to be a regimental wedding. Mess dress would be required for the first event and No. 1 Dress (Blues) for the second. It seemed to me that I had joined a very social organisation. Only twelve months previously I had been a schoolboy. Now I was commissioned, about to command a platoon of regular soldiers and was having to learn what were then the very formal social mores governing life in an officers' mess. It was a huge transition for a nineteen year old to make.

Every infantry officers' mess carries with it an impressive amount of silver – e.g. large centre pieces, cutlery, goblets, claret jugs and even snuff and cigar boxes. It is all owned by the regiments and the cost of its maintenance is borne by their officers. Its value runs into many hundreds of thousands of pounds and ours was routinely inspected and repaired by Garrards – but at a price commensurate with the company's standing. When a long dinner table is decorated with it, against the backdrop of a purple table runner and the regiment's Colours, the result is striking. At the mess dinner, I kept as low a profile as possible, but as part of 'after dinner entertainment', I rode a donkey along the mess corridors and through the dining and ante rooms, in competition with fellow subalterns. While I had thoroughly enjoyed the dinner, I took much more pleasure in the fact that I was now included in a remarkable military family.

The regimental wedding was held in Downpatrick Cathedral. Captain David

Thorogood was to marry a local young lady, Sue, and after the ceremony, the reception was held in the mess at Ballykinler. The reception was a perfect blend of military and Irish hospitality. It was a very happy, relaxed occasion. It was also a chance to get to know my fellow officers a little better and I was introduced to some incredibly kind, local civilians, who were to offer frequent and generous entertainment at weekends.

I had been posted to 'A' Company and was to command 3 Platoon. My company commander was a pleasant, urbane officer, Major David Page. I did not see a great deal of him, or his second-in-command, Captain Mike Duffy. In truth, for much of the time there was not a great deal for them to do. Platoons got on with training, which was largely conducted at a personal level, i.e. fitness, weapon handling, shooting and individual field craft skills. It rarely involved the hierarchy.

The previous commander of 3 Platoon was Paul McMillen. I had yet to meet him, because he was attending a course. On his return I shared a room with him. A ceiling to floor set of built-in wardrobes and cupboards divided the room and gave us a modicum of privacy. While that was welcome, it was of no great importance because we got on exceptionally well from the outset. Paul was a rotund, jolly officer, who saw the best in people and the funny side of virtually everything. On the downside, his jokes were dreadful and never improved. He was to become my closest friend and remained so until his untimely death in 2012.

My day-to-day work activity was guided by Sergeant Hutchinson and to a lesser extent, by my three experienced section commanders (corporals). It remains accepted that while new officers have earned their commissions, they still have a very great deal to learn, particularly with regard to man management. A week or so after my arrival, Sergeant Hutchinson took me to one side and said, 'Which of it is it to be, sir? Are you going to be a nice guy, or a proper bastard? It doesn't much matter to us, as long as you are consistent.'

I was a bit taken aback and told him I would try just to be myself and be consistent in that. I had the good sense to say that I already valued his advice and would continue to look to him for guidance. We got on well and a judicious nudge from him here and there, helped me feel increasingly confident as the weeks progressed. However, I had only scratched the surface and two incidents illustrate that and demonstrate the close, complex relationships that exist between young officers and their NCOs.

National Service was fast drawing to an end. In my platoon was the last national serviceman to serve in the battalion – 'Rubber Lips Driscoll'. He hated the Army and was truly the worst soldier I ever encountered. My platoon's barrack rooms were on the second floor of the soldiers' accommodation block. One morning I was to conduct a formal room inspection with every soldiers' kit immaculate and laid out on their beds, lockers and chairs. Private Driscoll was to leave the Army at the end of that week. There were only two or three days left and everyone, irrespective of rank, would be pleased to see the back of him. I moved slowly from one bed space to the next, occasionally praising some particularly well presented piece of kit, or ordering something that was not up to scratch to be 'shown again'. Sergeant Hutchinson moved one pace behind me. When we got to Driscoll's bed space it was obviously well below par. His best boots had hardly been touched. 'Is it worth it?' I thought to myself and moved to walk past. There was a cough from behind me.

Sergeant Hutchinson said, 'Driscoll – Mr Groves had a hard night in the mess and is not himself this morning. If he was he would have said what a fucking disgrace your boots are. You know I'm a sporting man. Open the window behind you and give me your boots. Now go and stand by the door. When I say "go" I will let go of your boots and you can run down the stairs and catch them!'

Needless to say Driscoll's boots suffered badly. He had to get them up to standard before his kit was accepted for his discharge. But that is not the point of the story. On the way back to the platoon office, Sergeant Hutchinson was unusually quiet. The office was empty and he closed the door before rounding on me. He was as angry as I ever saw him.

'Don't ever do that again. Don't ever let your men down. Don't ever let me down and above all, don't ever let yourself down. You're better than that!'

I was mortified, ashamed and abjectly sorry. It was a lesson I took to heart and whenever I have been faced with a difficult decision since, in any aspect of life, Sergeant Hutchinson comes to mind and still provides all the reinforcement needed.

On another occasion my senior corporal asked to see me formally. He communicated through Sergeant Hutchinson who advised that he should be present at the interview. I quickly agreed, but the corporal would have none of it.

'It's just you I want to see, sir.'

Sergeant Hutchinson was not best pleased, but had no choice but to leave us.

'What is it?' I asked.

'I'm having trouble with the wife, sir.'

The corporal was in his upper thirties. I was nineteen. He could, just about, have been my father! That sort of situation was not taught at Mons. It was a very personal problem and with the corporal's approval, I enlisted the help of the padre, who had access to a range of organisations able to provide support. Thankfully, the matter was satisfactorily resolved. Most interestingly, despite my young age and inexperience, it was his officer, not a fellow NCO, who the corporal trusted to help and keep the matter entirely confidential.

I had joined the battalion in the summer. Cricket and athletics were in progress and football was in prospect. I was a proficient wicketkeeper from all the time I had spent behind the stumps to my brother's bowling, so I was drafted into the battalion team in that position. Athletics had more interest. The battalion had a physical training instructor called Lance Corporal George Gooden. He could run 100 yards in 9.6 seconds and had made the UK's 4 × 100 yard relay squad. His efforts galvanised interest in athletics and the battalion formed a reasonable team, with myself as part of it. The team was entered for the Army Championships and although it did not make the finals, a solid athletic base had been established for the future.

Football is the soldiers' sport and there was a lot of interest in the new, young officer, who was supposed to be a reasonably good player. The trials for the new season went well for me and for that other newcomer, Private Taylor. We established ourselves in the team with myself in goal and Graham Taylor as a wing half. Again my sporting ability had assisted me to make an impression and that was invaluable to a young officer trying to make his way.

It is probably already clear that sport played a major part in battalion life. In fact, there was very little else to occupy soldiers in their free time. Ballykinler's remoteness meant that leaving the place required a considerable amount of planning and spare time. Cars were not commonly owned below the ranks of captain for officers and sergeant for enlisted men. Only two cars were owned by subalterns, a Mini-van and a kit car; for junior ranks car ownership was out of the question. Officers and sergeants had their messes for their entertainment. Junior ranks had to rely almost entirely on the NAAFI (Navy, Army, Air Force Institute) and an establishment called Sandes' Home, run by a Christian based charity, for drink, fast food and pub games. It was all they had day-to-day. At weekends the most adventurous of them might get to Belfast. More found diversion in Downpatrick, a small town that was much more local, but lack of funds kept many in barracks. For them, Ballykinler was not a popular posting.

All officers were required to hold a driving licence and so, for a short while, I was placed in the care of Colour Sergeant 'Blinky' Tranham, the Motor Transport Platoon's SNCO. He was a qualified testing officer and could award civilian licences. He arranged for certain of his JNCOs to instruct me before he tested me personally. The vehicle we used was called a 'Champ'. It was a totally over-engineered Jeep that was capable of fording rivers. Its engine was made by Rolls Royce. It did not have a synchro-mesh gear change and it was as heavy as lead to steer. However, I had been well taught and my test was going very well until we reached reversing. The roads around Ballykinler were often extremely narrow, some with high banks and hedges on either side. On one such road I was asked to stop just beyond a lane. Blinky was an old fashioned gentleman, with the manners to match.

'Would you mind reversing into the lane, sir?' I did as I was asked until there was a scrunch and a breaking of glass. 'I think you should pull forward, sir, and I'll see what's happened.'

Blinky got out, walked to the back of the Champ and proceeded to extricate three or four milk crates and their shattered contents from under the vehicle. He tossed the crates over a hedge. 'Bang goes my chances of passing,' I thought.

Blinky got back in. 'Bloody silly place to leave milk bottles,' he said and I got my pass certificate on return to barracks.

Ever since I had joined the battalion in the summer, the talk had been of a royal visit to be made in late October by Queen Elizabeth, The Queen Mother. The Army enjoys a close connection with the royal family and all regiments and corps have royal Colonels-in-Chief. We were lucky enough to have The Queen Mother as ours. Her visit would be my first experience of meeting a member of the royal family and there would be many in coming years. However, I was to become something of a 'Jonah' where such events were concerned; someone to be kept out of the way if things were to run smoothly. Whenever I met a member of the royal family, something untoward would happen and the first meeting with The Queen Mother set the precedent.

The planning for the royal visit was done in minute detail. The Queen Mother was to inspect a parade, meet some soldiers and their families, visit the warrant officers' and sergeants' mess and then lunch with the officers and their wives. Everything was rehearsed, including the lunch, which was overseen by a particularly pompous major, the President of the Mess Committee (PMC). The day of the Queen Mother's visit was grey, overcast and not very warm. Having completed the earlier part of the programme, the Queen Mother arrived at the officers' mess. She was to withdraw for a few minutes to a freshly decorated bedroom, which happened to be on the ground floor. My own room was the next beyond it. The PMC wanted to present each officer and wife individually, so we were gathered just outside the mess to be called forward at the appropriate time, into the impressively proportioned entrance hall where the presentations would be made.

The Colours had been carried on parade by two senior subalterns and had to be displayed in the dining room for the lunch. That posed no problem, however, once it had been done, the two officers still wore the ornate belts that had permitted them to carry the Colours efficiently.

'Get rid of these,' I was told and was given the two belts. I trotted off around the outside of the mess to go to my bedroom. Each of the rooms on that side of the building had two sash windows that came down to just above waist height. I thought that I would simply duck under the Queen Mother's windows, throw the belts in my own room and repeat the process on the way back. Having tossed the belts into my room, I ducked under the first of the royal windows, but I came up slightly too early at the second. I was face to face with Her Majesty. She looked surprised and then smiled. I made a hurried bow and returned to the mess entrance. As the junior subaltern, I was the last to be presented.

'Mr Groves, ma'am,' intoned the PMC. I came forward to bow and take the Queen Mother's hand.

'Ah, yes. We have met,' she said.

It was as priceless as it was gracious. The pompous PMC was thunderstruck. Magic moment.

As a newly commissioned second lieutenant, my monthly pay, after tax, national insurance and the like, was £37. From that I had to pay my mess bill, laundry and tailor's standing order. If I had to watch the pennies, it was much worse for the enlisted men. Soldiers were paid weekly, 'over the table'. Every week each company would send a subaltern to the pay office to collect and sign for the pay for his company. In 1963, a company's weekly pay, for more than one hundred men, amounted to just short of £1,000. The paying officer would be accompanied by a clerk form the Royal Army Pay Corps and two soldier witnesses, drawn at random from his company. The company would parade in front of a table with the officer and clerk seated and the witnesses looking on. Names would be called out and each NCO and soldier, in turn, would present his pay book. The clerk would state the amount to be paid and the officer would count it out, sign and return the book containing the money to the soldier, who would announce, 'Pay and pay book correct, sir'. Even at that time it was an archaic and time-consuming process, but virtually no soldiers had bank accounts, indeed most came from families who had never held a bank account. For their part, the subalterns fervently hoped that they had got things right, because if they were short of funds at the end of the parade, they had to make good the shortfall from their own, not very deep, pockets.

I quickly learned that the Army was infinitely capable of surprises. I was to be sent on my first external course, an adventurous training one that would qualify me to instruct on rock climbing at an elementary level. The first surprise was that I was to report to the Royal Army Education Corps' (RAEC) centre in London SE9 and it seemed highly unlikely that there were any rock faces in that part of the capital. When I got there I found that the RAEC had taken over Eltham Palace and its officers were living in the lap of luxury. Not only did the palace have a splendid medieval great hall and stunning gardens, but those assets were completely eclipsed by 1930 additions built for the Courtauld family. Each of the modern rooms was a masterpiece of Art Deco design and even the standard Army issue tables, chairs and soft furnishings did nothing to detract from the magnificently crafted walls, floors and ceilings that surrounded them. Unbelievably, this architectural jewel was now an officers' mess. The beneficiaries of all this were mainly long-in-the-tooth 'schoolies', who did not appreciate having their tranquil, well-ordered lives interrupted by a gang of noisy, young officers who clomped around their tasteful, elegant mess in boots! My bedroom was en suite, unheard of in the Army at the time. The first morning I got out of my bed, my bare feet touched … underfloor heating!! Oh, wriggle those toes – I had died and gone to heaven!

The first week of the course was conducted from Eltham. We travelled daily to somewhere in Kent that had ideal, nursery rock faces and we mastered the basic techniques there. The second week was held in much less salubrious surroundings – a 'wriggly tin' (corrugated iron) training camp in north Wales. Any heating, let alone the underfloor variety, would have been nice. Outdoors, Snowdonia's rock faces were in a different league from those in Kent and certainly concentrated the mind. Overall it was a good and enjoyable course, but although I did a lot of adventurous training later in my career, I was never to use my hard earned qualification.

On my return to Ballykinler I found that the battalion was to train at Stamford in Norfolk. Training at individual and platoon level provides the basics, but it was vitally important to progress to conduct exercises at company and battalion level, which required sophisticated ranges and the sort of large training area that Northern Ireland could not offer. To do that we were to travel to East Anglia. To get there was a logistical nightmare. First, some 600 officers and men, their weapons and equipment, were moved by road to Belfast docks. Then came the overnight ferry crossing to Heysham, to link with a dedicated troop train to Euston. The real fun started then. The battalion got out of its train in London and formed up in companies, on the platform, under their sergeant majors. Ours was CSM Martin Franks. It would be a profound understatement to say that he was a larger than life character.

The first thing he did was to get rid of his officers. He commandeered four taxis and bundled us into them with our kit, saying, 'See you at Liverpool Street, gentlemen.' With that he turned his attention to getting just over one hundred men and their gear through the underground system to the eastern side of London. As our taxis left Euston we could see our company, in step, gradually disappearing into the maw of the Northern line. We officers arrived safely at Liverpool Street and were waiting, with a mound of kit, close to the underground's principal exit.

After quite some time, from the depths, we heard, 'Left, right, left, right. If you want to join us, sir, pick up the step, otherwise stand back.'

Bit by bit the company emerged from the underground's staircase, three abreast, still reasonably in step and with CSM Franks enjoying himself hugely. The journey was not over yet. Another dedicated troop train took us to East Anglia and two weeks later, we did the whole thing in reverse.

I thoroughly enjoyed training my platoon with the help and advice of all my NCOs. The training at Stamford was at a level that I had not experienced before and I probably benefitted from it more than anyone else in my platoon. Over time we were developing trust in one another. I was growing in confidence, both of my ability to command and of our collective effectiveness. All of that was excellent preparation for a platoon test exercise, held in Northern Ireland, some nine to ten months after my arrival. The exercise was staged over three days to put our fitness, shooting and tactical abilities under detailed scrutiny. The first day involved a very long approach march of about 30 miles, carrying plus of 40 pounds of weaponry and equipment. Accurate navigation was the key if the marching was to be kept to the 30 or so miles. At the end of the day, our shooting was tested before we were allowed to establish overnight bivouacs. The next day brought river crossings, attacks and patrols. Although the distance to be marched that day was probably halved, the second night was spent patrolling against other platoons, so competition was extremely high. The exercise concluded on the ranges at Ballykinler and involved physically demanding fire and movement practices. It really stretched men who had had very little sleep and more than 50 miles marching under their belts. My platoon did well. We were placed highly and were extremely pleased with ourselves. It did my reputation a great deal of good and I began to think about converting my short service commission into a regular one and the full career that that would bring.

Sport continued to be my principal pastime. The football season was drawing to an end. At the time there were not enough military units in Northern Ireland to form a league, so we played against local civilian sides. Some away games were in staunchly republican areas. At one, in Newry, small children were encouraged to throw stones at any soldier near a touch line. As the goalkeeper, I got the lion's share!

Athletics was to be the highlight that year. Lance Corporal Gooden and I were entered for the Army's individual championships in the 100 and 220 yard events. He won both and I recorded a 10.0 seconds time for the 100 yards, but it was heavily wind assisted, so it did not count. I had better luck when it came to the inter-unit championships. This time the battalion team had made the finals. I was part of the 4 × 110 yards relay team that won the final in a time of 43.6 seconds. It is a record that stands to this day, but only because the following year the measurement for the race became metric.

Ballykinler was an ideal place for me to settle into my military career. The pace of life was not particularly demanding. I felt I had matured considerably, grown in confidence and developed as a platoon commander. I had made friends and had a pleasant social life. But change was in the offing. Shortly after my arrival in Northern Ireland, the Commanding Officer had brought the battalion together to inform us that in the summer of 1964, we were to be posted to Berlin, about as far removed from the rural remoteness of Ballykinler as it was possible to be. Real excitement beckoned.

CHAPTER FOUR
BERLIN – A SUBALTERN'S LIFE BEHIND THE IRON CURTAIN

THE MOVE TO Berlin was by air. There were three air corridors leading to Berlin and aircraft were required to fly at only 10,000 feet when traversing East German air space. From that height it was easy to see the border fence, sterile strip and guard towers of the inner German border that separated the Deutshe Demokratische Republik (DDR) from the Federal Republic. As we came in to land at RAF Gatow in West Berlin, we were amazed to see that only a road separated the airfield from the border fence with the DDR. It brought home how cheek by jowl we were to be with the hostile powers that surrounded us.

West Berlin was situated just over 100 miles from the West German border and only 70 miles from Poland. At the end of World War II, the allied powers divided the city into four sectors and a line was drawn roughly north to south through its centre. To the east was the large Soviet sector. The western half was divided into three smaller sectors, the French in the north; the British in the centre; and the Americans to the south.

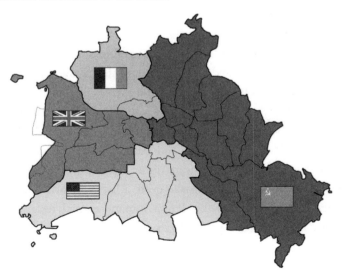

In July 1964, my battalion arrived to become part of the British Berlin Infantry Brigade. The construction of the Wall – begun on 13 August 1961 – had been completed and a vast wire fence barrier, similar to the one that separated East from West Germany, now enclosed the outer perimeters of the three western sectors. West Berlin was an island of democratic freedom, but it was also a sealed, besieged city.

The three East Anglian Regiments had only been in existence since the late 1950s, but the Army had determined that the future for the Infantry was the creation of still larger regiments. As a consequence, only weeks after arrival in Berlin, another organisational change meant the East Anglian Regiments were to form a new Royal Anglian Regiment and we were to become its 3rd Battalion (but still the Pompadours). That meant a huge effort on drill had to be made to be ready for the formation parade that was to be held on 1 September 1964. Parades were an important part of service in Berlin. Every year the British held a Queen's Birthday parade, different, but as impressive as the Trooping of the Colour in London. The British garrison also combined with the French and Americans in an even larger Allied Forces parade. These very public demonstrations of military capability were designed to reassure West Berliners of the allied commitment to their well-being and democratic way of life.

Relationships between East and West were at a low ebb, so much so that readiness for war was vital. Whether deployed in Berlin or West Germany, all British troops were at a constant two hours' notice to deploy from their barracks, ready and equipped to fight. That readiness was exercised regularly and woe betide any who failed to meet the stiff deployment requirements.

However, despite the tense political and military atmosphere, Berlin was a city to be enjoyed. I was fortunate to have joined the battalion at a time when it had a particularly lively and talented bunch of subalterns. Andrew Styles, Cliff Brock, Paul McMillen, Trevor and Jeremy Veitch, Brian Harrington-Spier, David Norbury, Tony Winton, John Hart, Dick Tewkesbury, John Goodfellow and Pat Shervington make quite a list, which is not exhaustive. Most became lifelong friends, who remain in regular touch today. They knew how to enjoy their soldiering and life in general. Together we were keen to grasp all Berlin had to offer.

Service men and women received an overseas allowance that improved our spending power. West Berlin was a showpiece for western democracy and there was a lot to see and do. If we wore uniform we could visit East Berlin and did so regularly to shop for Meissen china, communist paraphernalia and cameras, and to go to restaurants and the opera, where our scarlet, mess kit uniforms drew a lot of attention. In the West, the British Council provided an easy and convenient place to meet German civilians. The British military headquarters was splendidly located in that built for the 1936 Olympics. It was surrounded by unparalleled sporting facilities that were retained for British use. There was a reasonably good NAAFI shopping and cinema complex; an officers' club and the opportunity to use the corresponding facilities in the French and American sectors. The British did not have their own television station, however, the British Forces Broadcasting Service provided excellent radio programmes, which were listened to widely by all three allied nations and, more importantly, Germans on either side of the Wall.

Added to all that, alcohol was cheap and there always seemed to be a party in progress somewhere or other. With improved income, car ownership became a real possibility and the price of petrol was also discounted for the military. I bought a new, green Mini and it became my pride and joy. For bachelor officers the biggest boost to their social life was the presence of female Foreign Office secretaries – no further explanation necessary.

While the social scene was hugely enjoyable, the reality of our situation was never far away. A few months into the posting, I had decided to have a pre-dinner beer with a fellow officer at a local German pub. We were walking back to barracks when a two-man Ferret scout car lurched around a corner, travelling far too fast, with the commander standing on the back deck hanging onto the small turret. We were close to the barracks, so ran to find out why the vehicle had been driven so recklessly.

The Ferret was parked just inside the barrack gates. The commander of the vehicle, a Reconnaissance (Recce) Platoon corporal, had dismounted and was helping a dishevelled, young, civilian man out of the turret. That man was followed by another and then a young woman. We demanded an explanation.

'They've just come through the wire, sir,' the corporal said.

He explained that he had been conducting a routine border patrol, when he had spotted the three East Germans crawling across the wide, cleared strip of earth that was on the DDR side of the border fence. The three would have been in plain view of two sentries in a Volksarmee watch tower, but the sentries' attention was firmly on the Ferret scout car. The corporal and his driver pretended that their vehicle had broken down and made as if they were working on it. Very, very fortunately that ruse retained the East German sentries' attention. The three escapees got to the wire and cut their way through. The two British soldiers got back into the scout car and drove as fast as they could to position it between the escapees and the watch tower. One by one they loaded the three youngsters into the turret – no mean feat as it was designed for just one person. The vehicle then raced back to barracks. The three young East Germans were all weeping with a combination of shock and elation. They could not thank us enough and constantly mentioned 'freiheit' (freedom). Escape to the West had affected them profoundly and the episode was one of my life's defining moments. I knew, with absolute certainty, that the regimes that we were confronting were truly evil and had to be resisted. It gave a real sense of purpose to my soldiering, a purpose that was to stay with me for the next thirty years.

Berlin meant a lot of guard duties for the soldiers. Officers had other duties to perform. Other than in Berlin, most Army units have a subaltern act as duty officer every day. His/her duties entail visiting meals, conducting guard mounting, checking stores, visiting prisoners and supervising the various parades that punctuate the working day. In Berlin, two more subalterns were required for other daily duties.

The first was a tester. While the battalion was at two hours' notice to deploy, a platoon, the Alert Platoon, was always maintained at ten minutes' notice to be out of barracks, ready to fight. It meant that for the twenty-four hours that its duty lasted, all the soldiers of the nominated Alert Platoon had their weapons, ammunition and fighting order with them wherever they went. They slept in their clothes, not even removing their boots. The platoon's transport was loaded with war scales of ammunition and other fighting equipment. When a warning siren sounded, the Alert Platoon commander had five minutes to get to the adjutant's office to receive his deployment orders. He had five more minutes to get back to his platoon, brief them and go. The performance of all Alert Platoons was closely monitored, something that was taken very, very seriously.

Because the military situation was so tense, there had to be a secure alert system. That took the form of an inconspicuous, grey telephone, located in an austere, sparsely furnished, narrow room in battalion headquarters. The telephone was manned by a third duty subaltern, twenty-four hours a day, every day of the year. It was connected to the British headquarters. The system was tested daily when a disembodied voice from the headquarters gave half of a two-word code and the recipient would respond with the other. Once the code words were successfully exchanged, the job was done for another twenty-four hours, unless the balloon went up, in which case the system would be used in earnest, so despite it being an intensely boring duty, it was an important one.

I was unlucky enough to be assigned this duty on Christmas Day 1964. I took over at nine in the morning with plenty of books to see me through the interminable twenty-four hours that lay ahead. The highlight of the day was my Christmas lunch, which was brought to me to eat alone in that soulless room. After I had eaten the meal, there was nothing much to look forward to until the next one and the test call that was generally made around eight o'clock in the evening. However, in the mid-afternoon there was a lot of noise outside my door. In the corridor were six or seven soldiers of my platoon. Obviously, they had had a lot to drink and were carrying a couple of crates of beer with them. 'You need a drink, sir.' I explained that I was forbidden alcohol, but they ignored that and flooded into the room. Tops came off bottles and one was thrust into my hand.

and I told myself that
first beer and the oth
The boys eventually
evening meal and th
the minutes away u
response. I could no
in an instant.

I woke knowing
light in the corrido
my door.

'Who's there?'

Still half asleep
said more loudly
whispering and
battalion headqu

... it was a kind thought

the real estate had to be shared with the m
was sufficiently large for platoons to co
city. However, a large part of Berlin w
Military exercises in Berlin w
amongst the NCOs and men
Amazingly, the story
Rover was called and
required to rem
lady was mu
laughing
I

shoulders. Barefooted I opened the door. In front of me was a very, very large woman, hopelessly drunk, wearing a home-made sash. On it was scrawled 'Happy Christmas'. It was all she was wearing!

She looked past me. Seeing my bed, she pushed me aside with the power of an international rugby prop forward – took three or four paces into the room and flopped, face down on the bed and immediately started snoring. I stared at an enormous bottom that pointed skywards. What on earth was I to do? Mine was surely destined to be the shortest commissioned career in history. But desperate times bring desperate measures. Fortunately, the duty policeman that day had previously been a member of my platoon. His name was Lance Corporal Lyons. I rang the guardroom and told the guard commander to wake Lance Corporal Lyons and have him ring me back. A few minutes later, he rang and I asked him to come around to my room, as I had a problem and needed his help. Meanwhile the snoring persisted. The lady was impervious to any attempt to wake her.

I heard Lance Corporal Lyons' boots clomp past my window, up the steps of the building, and stamp to a halt outside my door. I was now fully dressed – shirt tucked in, tie aligned, tunic on, Sam Browne belt in place, polished shoes on my feet and hat properly adjusted. The room was so narrow that I had to stand adjacent to the large bottom. Lance Corporal Lyons knocked. 'Come in,' I said. The door opened. He took a pace forward and his right hand rose to salute, but it faltered at about shoulder height. His amazed expression told me that he had misinterpreted the situation.

'Sorry, sir. I'll come back later!'

grabbed him and explained what had happened. When he stopped he said that he would go and get some of the guard, because the ch more than a two-man lift. Six of them and three blankets were ve her. Apparently, she was well known, so the duty Land she was delivered home safely.

did not reach the 'powers that be', but my standing went up enormously.

re limited by the confines of an international forest and lake. The forest, the Grunewald, duct some worthwhile training, but ny civilians who wanted to take advantage of its comparative peace and green expanses. My first experience of exercising in the Grunewald brought an unusual encounter and another lack of clothes.

My platoon was advancing tactically through the forest. We moved carefully and made little noise. I was travelling behind the leading section when I saw its commander crawl towards me and tap his shoulder. The signal indicated that he wanted his officer – officers' badges of rank are on the shoulders of their uniforms, thus the signal. I crawled to him and, whispering, asked him what he wanted.

'There are a lot of naked women running around just in front of us, sir. But don't get your hopes up. They all need ironing!'

The explanation was simple. In the 1920s and 30s, Berlin developed a reputation for libertine behaviour, even decadence, and if there was any connection at all, the naturalist movement was at the respectable end of that spectrum. The movement was still strong in the 1960s and my platoon had stumbled across a number of its long standing, possibly even its founding, members. We circled past them.

If that was mildly amusing, then the Berlin Brigade Platoon Battle Tests were not. Every year, in the depths of winter, a test exercise was organised that required the participation of every rifle platoon of the three infantry battalions in the brigade – twenty-seven platoons in all. Neutral umpires were brought in from the British Army of the Rhine (BAOR) to judge the tests and for three days and nights, every phase of war, plus fitness, stamina, field craft and marksmanship skills were scrutinised. The results were made very public, which heightened the already fierce inter-battalion rivalry. The pre-training was as rigorous as the tests and for a work-up period of about three months, the competing platoons' efforts were directed at very little else. My battalion was the most recently joined unit, thus we were at a severe disadvantage (the others had the experience of the previous year), but our platoons performed creditably. I was still amongst the most junior subalterns and was judged to have done well. That comparative success was the catalyst that decided me to apply for a regular commission.

I was due some leave and elected to drive the 600 miles home from Berlin. Like all allied travellers using the road corridor to West Germany, I had to obtain a travel document and be briefed by the Royal Military Police (RMP) on the precise route through the DDR and what to do in case of trouble. The journey started at Checkpoint Bravo, at the south west extremity of West Berlin, where the document was presented and British travellers were booked out by the RMP. Travellers then left to travel only a short distance to a Soviet checkpoint (allied forces did not recognise the East German authorities and dealt only with the Soviets). At that checkpoint all drivers had to approach a Soviet sentry, exchange salutes, and present the document. The sentry would inspect the document before returning it and indicating that it should be taken to a hut for more formal scrutiny. Inside the hut, travellers were required to knock on a boarded up window. A wooden panel was raised about six inches and a hand would be extended to accept the document. The panel was then closed and a wait began. If you were lucky it could only be a few seconds. If you were not, twenty minutes of kicking your heels was not uncommon, made worse by the hut's unpleasant odour of foul cigarettes and unwashed soldiery. The document was eventually returned with a Soviet stamp on it. Barely a word was exchanged. This was the front line of the 'Cold War' and the prevailing atmosphere was frigid. Outside the document had to be shown to the sentry before travellers drove the 100 miles towards West Germany, along potholed, largely deserted autobahns. Speed limits had to be scrupulously observed, so the journey took two hours. About a mile from the West German border was another Soviet checkpoint and the stately dance with the document had to be repeated, before booking out with the RMP at Checkpoint Alpha, at Helmstedt and release into the western world.

I was keen to see my family and Sue again. We had kept in contact by letter throughout my time in Northern Ireland and Berlin, though she was much better at writing than I was. However, she had met another man while studying at Loughborough. When we did meet, she told me that she had become engaged and our boyfriend/girlfriend relationship was at an end. That was upsetting, but I had had occasional, short friendships with girls in Berlin, so it was understandable that there should be a parting of the ways. She had met me at my parents' home, now in Sudbury, Suffolk. I drove her back to Loughborough, but managed to crash my car on the way there. We came to a shuddering halt in a pub forecourt, with the car upended on the driver's side and Sue on top of me. The publican had obviously been waiting for such an incident to happen for years. He appeared instantly, like a genie out of a bottle.

He pulled Sue out of the car through her window, scraping her back in the process. He then led her across the forecourt's sharp stone chippings to the pub. She had lost her shoes in the impact and still lying on my side in the car, I saw her dance across the chippings like the 'sugar plum fairy'. Even though she was in a confused state, I thought it better not to laugh as the incident was entirely my fault. It did nothing to endear me to her and we said our goodbyes, but agreed to remain friends.

Just how much West Berlin was a hostage to fortune during the Cold War is difficult to over-emphasise. From June 1948 until April 1949, the Soviets closed the land routes to Berlin and the city had to be resupplied by air. Ever since, large stocks of food and other essential supplies were maintained in the city as a contingency against another crippling blockade.

In the spring of 1966 East/West relationships deteriorated badly. The West German government announced its intention to hold a parliamentary session of the Bundestag in West Berlin. That was contentious and promoted a furious reaction from the Soviets and East Germans. The communist authorities' dramatic response was to close the road transit routes to the city, causing huge tailbacks and delays that frequently lasted the better part of a day or more. Train and air routes continued to function, but the interdiction of the road corridors raised fears of another blockade.

To increase the pressure on West Berlin's citizens still further, the Soviet and DDR air forces flew jet fighters over the city for days on end, breaking the sound barrier frequently. The huge bangs and overpressure caused shop windows to shatter, provoking enormous anxiety amongst the civilian population.

Apart from inter-governmental protests, the allied response was to send military vehicles along the road corridors to exercise their right of access to West Germany and to the city. In order to avoid armed confrontation, soft skinned vehicles – Land Rovers and DKW Jeeps – were sent, crewed only by an officer and a driver. The probes met with limited success, but they were sufficient to demonstrate to stranded western travellers that the allies were with them and pressing for the blockages to be lifted.

The Soviet excuse for closing the road routes was that large scale military exercises were being held and that the exercises crossed the autobahns, thus it was unsafe for civilian traffic to be in transit. To add credence to this claim the Soviets and DDR authorities used tanks and armoured vehicles to block the routes. Once a route was blocked, some hours would pass before the armour would move aside and traffic would be allowed to flow, but it was never certain how far it would be permitted to go before another blockage was imposed.

The city was extremely tense and the buzzing by communist fighter jets was loathed above all else. On the 6 April 1966, a Soviet fighter spiralled out of control into the Stossensee Lake in the British sector of the city. To their eternal credit the pilot and navigator stayed with the plane until it crashed, thus preventing mass casualties that certainly would have resulted from an impact in a built-up area. Immediately, the West Berlin emergency services were on hand, but as the occupying power with jurisdiction, the British took control of the situation. First the crash site had to be secured, so an alert platoon was deployed to cordon off the area.

The Soviets were extremely keen to conduct the recovery of the plane and its crew, but their recovery vehicles were stopped and turned back to the eastern side of the city. However, the Soviets had every right to send observers to the scene and they did so extremely quickly and in significant number. They established themselves inside the British cordon and were equipped with large telescopes and huge binoculars, even though they were only a matter of a few yards from where the recovery was being conducted. Within a day they had brought in large barges and cranes to remove the wreckage, but just as quickly Royal Navy divers were there to locate what was left of the plane and to direct the operation in the Stossensee's very muddy waters. Visibility in the water was almost nil.

Within three days the bodies of the crew were recovered and repatriated. The main part of the fuselage was also quickly located, but the engines, radar and radios proved much more difficult to find – or so the British claimed. As each piece of wreckage was located, the cranes would lift it and place it in one of the East German barges for them to return it to the DDR. The Soviets were desperate to recover every last piece of the wreckage because the plane was a Yak-28P, their latest all-weather fighter, containing their most sophisticated technology. To lose that to the West would have been a disaster for them.

It took several more weeks for the missing components to be found. However, when they were located and raised to the surface, they were in pristine condition. It transpired that despite the close watch kept by the Soviet observers, the naval divers had secured ring bolts to the bottom of some of the barges and had raised the engines, radars and radios to secure them under the barges' keels. When the barges made the journey back to the DDR they passed the British yacht club, on an adjacent bank. The key components were released from the barges and using some sort of buoyancy aid, they were floated to the yacht club, which happened to be only a few hundred yards from RAF Gatow. British aviation experts had been flown into the air station and there they examined the latest Soviet technology at leisure and in minute detail. Having gleaned all they could, it was back to the naval divers to return the vital components to the crash site.

It had been a brilliantly executed operation. The Soviets were as amazed as they were upset, at the loss of their technology. Everyone was staggered when the gleaming engines and other equipment, broke the water's murky surface. How on earth the divers had managed to do what they did, without discovery, was little short of a miracle – the stuff of a Bond movie.

The Bundestag met in West Berlin on the 7 April 1966, despite the DDR prohibiting Bundestag members from passing through its territory (the western politicians made the journey by air). After the parliamentary session, the situation gradually eased and a degree of normality was re-established.

Our soldiers' routine centred on the many guard duties. However, young officers were fortunate to have some more exciting tasks. These included 'flag' tours into East Berlin; Spandau prison guards; and commanding the military train to and from West Germany.

Providing the guard at Spandau prison was interesting because it held the surviving Nazi hierarchy convicted at the Nuremburg trials. During the battalion's tour in Berlin, Hess, von Schirach and Speer were in residence. The four occupying powers took turns, a month at a time, to provide the military guard that secured the prison's perimeter. Civilian warders, drawn from the four powers, looked after the prisoners.

The prison's crumbling, red-brick edifice squatted sinister and unlovely, adjacent to a British barracks. Even in fine weather there was an eerie malevolence within the gaol's walls. In November 1964, I was the guard commander there. My platoon provided the sentries. Six guard towers were built into the perimeter walls and overlooked a series of barbed-wire topped, mesh fences that prevented any close approach. The guard commander was tasked to patrol within the walls at frequent, but irregular intervals. As he approached, in order to demonstrate that they were alert, the tower sentries would report to him that their posts were correct.

During the night of my duty, the air was chill and a mist had descended. The orange glow thrown by the sodium security lighting made the interior grounds of the prison a ghostly place, an ideal location for a Hammer horror movie. The intensely disconcerting atmosphere was made worse by the frequent falls of masonry from the dilapidated jail. Only the cell block had been maintained since the end of World War II. The garden was worst and utterly spooky after dark. Tall bushes and sunflowers bordered the path and, in some places, shrouded the nearest sentry's view into the prison's confines. The garden was never silent. Constant creaks and rustlings added to the eeriness.

In the early hours of that morning I was conducting a patrol and approached the tower that overlooked the garden. My shoes scuffed the fallen leaves on the path. Suddenly, I sensed someone behind me. I stopped. The noise behind me stopped. I walked on. The noise began again. I stopped. It stopped. Alarmed I drew my pistol and walked forward a few more yards. Again the noise commenced. I moved quietly into the cover of some sunflowers, but continued to move my feet. The noise kept coming. Without cover from the nearby sentry I was going to have to confront the intruder alone. I sprang out, my cocked pistol clasped in both hands, absolutely ready to despatch my pursuer. There in the middle of the path, about 10 yards back, was a hedgehog! Thank God for the sunflowers. I would never have lived that down.

Each of the occupying powers provided a prison director. For the Americans, French and Soviets, they were serving colonels. The British director was a retired officer – very experienced and cheaper for the tax payer to fund. The directors lunched together, reasonably regularly, in an officers' mess that was immediately outside the prison walls. I happened to be the guard commander on the day of a British sponsored lunch and was invited to join the directors. As it was a formal occasion, I was wearing service dress, which has lapel badges. The 3rd Battalion proudly had an imperial eagle as the centrepiece of its badge. The French director recognised it immediately and asked about its history. It was embarrassing for us both as I explained that the eagle was captured from the French at the Battle of Salamanca in 1812 by a forebearer regiment. While everyone else was prepared to let the matter rest, the American director kept teasing his French counterpart over and over again. Finally, to help the unfortunate French officer, I reminded the American colonel that the White House is not the original building. The first was made of wood and another of our forebearer regiments had burnt it during the American War of Independence. End of teasing.

Up to that time the Soviet colonel had only spoken through his interpreter. He got up from the table and left the room, to return a few minutes later with a bottle of vodka of unknown provenance. He gave me the bottle and shook my hand. 'Thank goodness there are some friends in this room,' he said in perfect English and, winking, sat down to continue his meal.

'Flag' tours in the eastern part of the city were always exciting. All four occupying powers had right of access to any part of the city and exercised that right on a daily basis. For the British that meant that an officer, or officers, would be collected by a specially trained Royal Corps of Transport driver, in what was then, a very new and powerful car, the *VW Variant* that could easily out-perform anything the East German Volkspolizei had (they were tasked to shadow us). A member of the intelligence staff would brief us and the task was normally to snoop around Soviet or Volksarmee barracks to see if we could detect anything unusual, or take pictures of their equipment. The most enjoyable bit was trying, and invariably succeeding, to lose the Volkspolizei tail.

Trying to cover major military parades in East Berlin was altogether more challenging. Teams of officers would be sent through Checkpoint Charlie into the eastern part of the city in the early hours of the mornings of the parades. I was part of one such team of three officers to help cover the May Day parade in 1965. After a day's training on a sophisticated and, ironically, East German *Zeiss* camera, we were carefully briefed about what pictures were required and of what equipment. Our driver dropped us off on the parade's exit route, at the eastern extremity of the city and then returned to the west. We had no means of communication. Our orders were to be as inconspicuous as possible and to make our way, on foot, back through Checkpoint Charlie when we had completed our mission. That was easier said than done as we were dressed in full uniform (civilian clothes would have made us spies – not a good idea). We stood out like sore thumbs, but the crowd was sparse on that part of the exit route, which was remote from the city centre, the saluting dais and the dignitaries in Alexanderplatz.

Eventually we got our pictures and began the journey of several miles towards

Checkpoint Charlie. At first all went well, but as we got closer to the centre, the crowds became much more numerous, with the Volkspolizei increasingly in evidence. They had no authority over us, but past experience was that they would try to take cameras from western military teams, expose the film, return the cameras and suffer the political protests that would inevitably follow.

We were spotted by some Volkspolizei. We had a reasonable start and dodged down a series of side streets. Eventually, we emerged onto a main road, filled with civilians carrying hundreds of red banners. This was part of the workers' parade that followed the military one. We had little choice but to dive into the parade, to put it between us and the Volkspolizei. Our muttered 'Entschuldigungs' (excuse me) were received with some surprise and curiosity, but no antagonism. Almost immediately the parade began to move with us in the middle of it. We went with it for a short while, but reckoning that someone would give us away sooner rather than later, we said our polite 'Auf wiedersehens' and exited. By now the crowd lining the route was several deep and we had to shoulder our way through to reach another side road. We took a circuitous route to Checkpoint Charlie and, amazingly, had no further problems. We were one of very few teams that managed to keep their camera and film intact.

Railway transport was important and the allies all ran their own trains for the routine movement of personnel and to maintain their garrison's supplies. The British Berlin passenger train was arguably the most famous of the allied trains. Other than on Christmas Day, the train ran daily from Charlottenburg in West Berlin to Braunschweig in West Germany and returned the same day. The train was staffed by an officer, who was assisted by a warrant officer, a specialist in railways, and a SNCO Russian interpreter. It had an armed guard of about six soldiers to prevent any unauthorised people boarding. In addition to the normal passenger carriages, all with large, metal Union Jacks bolted to their sides, there was a staff coach that provided accommodation, a radio cabin and food for a prolonged period. The train's pièce de resistance was its dining car that served a three-course lunch and dinner, but it was most renowned for its high tea.

Like the road route, the process for transiting East Germany was a complex one. The doors of the train were sealed by the guards before leaving Charlottenburg in the morning and an East German engine, with a trustee crew, would be attached to take the train across the border to Potsdam. There another smoking leviathan of an engine would be attached for the journey to Marienborn, a bleak station a few miles from the border with West Germany. Each passenger and the train's staff had the same travel documentation that road travellers had and all documents had to be inspected.

On arrival at Marienborn, the officer, warrant officer and interpreter would dismount, form up and march down the platform to meet a Soviet officer. The Soviets held all the cards and if it was raining it was not unknown for a Soviet officer to meet the British party at the edge of the station's canopy, so that the British had to stand in the rain. The travel documents would be handed over to be taken to an office for scrutiny and stamping. As with the autobahn, there was no telling how long this procedure would take and the British party would be left standing where they had halted. While the stately dance involving the documents was going on, the engine would be de-coupled and another locomotive, with a trustee crew, would be attached for the short journey to Braunschweig. With the documents returned the three British soldiers would remount their train and the journey could be completed, only for the entire performance to be repeated when the train returned to Berlin in the evening.

On one occasion the engine that pulled the British train through East Germany had small crossed flags attached to its front – those of the Soviet Union and the DDR – and a formal protest was deemed to be necessary. By chance I was to be the train commander on the following day. I was called to the British commandant's office with my interpreter and we were briefed personally by the General. He impressed on us that this was an inter-governmental protest, albeit at an initial low level. Much depended on how the protest was received as to what the British government would do next. He handed me a paper with the formal protest written on it. The interpreter got the Russian version. We were asked to read them before being taken to the train.

We felt very important when we got out at Marienborn. We marched towards the waiting Soviet officer, halted and I saluted. He returned the salute and held out his hand expecting the travel documents to be handed over. Instead I took out the protest and read it in full, in English. My interpreter then read the Russian translation. I proffered the protest and the Soviet said something. I turned to my interpreter.

'He says he can't accept it. He doesn't have the necessary authority.'

That was certainly not in the script. In desperation I went for a long shot. I said, 'Every Army has a report form for a duty like yours. Put the protest in that.' The interpreter translated. Not a muscle moved in my opposite number's face. Slowly, he stretched out his hand and took the paper. Before he had a chance to do anything else, my warrant officer dumped the box of travel documents on top of it so he had no chance of returning it. Mission completed, but it felt like a close run thing.

But not all encounters with the Soviets at Marienborn were so steely eyed. In July 1966, England had won the football World Cup against West Germany and I was the officer in charge of the passenger train a few days later. The formal handing over and scrutiny of documents had taken place and the British party were about to leave to board the train. As I saluted, the Soviet officer said something that departed from the normal script.

'What did he say?' I asked my interpreter. He said, 'We won.'

I did not understand so I asked for an explanation. A big grin crossed the Soviet's face. He raised a hand above his head and shook it as if waving a flag. 'Russian linesman,' he said.

Football fans among the readers will know that extra time was played in the final, during which the ball had hit the West German bar and bounced down. The English were convinced that it had crossed the line and a goal had been scored. The Germans were equally certain all the ball was not over the line and the score was still 2-2. The referee seemed uncertain until he spotted the Russian linesman with his flag raised. A goal was given. 'We won' now made a lot of sense.

Limitations on the scope for training in Berlin meant that twice a year, all three infantry battalions left the city, in rotation, to train in the British zone of West Germany. 'The Zone' had two large training areas controlled and used extensively by the British. One, some 50 miles south of Hamburg, was at Soltau. There the infantry would be joined by tanks, artillery, engineers and helicopters to conduct combined arms training. Soltau was large enough to permit tactical manoeuvre at battalion level and blank ammunition and pyrotechnics were used extensively to create realism. That level of training was entirely new to me. I enjoyed the complex inter-actions that it brought, but there was a lot to learn.

The second area was at Sennelager, where live firing exercises were conducted. The British Army has always been particularly good at live fire training. It is as close to combat as it is possible to get. Young soldiers experience the crack of live rounds passing close to them, fired by their colleagues, as they run, dodge, dive and take cover, only for them to fire as their colleagues do the manoeuvring. Their adrenaline surge is heightened by the smell of burnt cordite and the acrid fumes of the chemical smoke that helps cover their movement. The platoon's specialist weapons were also fired and thrown. Soldiers were never very keen to fire shoulder controlled anti-tank weapons because the back blast was decidedly unpleasant, covering the firing teams with dirt, while the over pressure cleared their sinuses. Hand grenades were even less popular. They are heavy and contain a reasonable amount of explosive, thus grenades detonate quite close to the thrower, even after a decent throw. If a grenade fails to explode, then it falls to the officer conducting the practice, to leave the safety of the shelter bays, to place an explosive charge on the failed grenade, ignite the fuse and withdraw. The grenade could go off at any moment, so a reasonable amount of time is given before undertaking this task. The first time that I was called upon to do it, I felt extremely exposed and glad when it was all over.

Once live fire has been mastered at individual, section and platoon level, the exercises become larger and Sennelager could cater for company exercises, with the infantry able to be supported by fire from mortars, anti-tank guns, medium machine guns, artillery and light tanks. The noise was tremendous, the heart rate high, and the learning curve vertical.

Even Sennelager could have its humorous moments. On my first visit there my company bivouacked in a tented camp in woods. Despite living in the field, much of normal routine was maintained. Muster parade was held every morning to ensure soldiers were properly turned out and with all the equipment they would need for the day's training. One morning, just as the officers were about to join the parade, I felt a call of nature that could not be denied. I disappeared over a small mound to where the officers' deep latrine toilet had been dug. Meanwhile, the Company Sergeant Major, now WO2 John de Bretton Gordon (DBG), wanted to know why I was absent from the parade. One of my fellow subalterns attempted to explain. DBG was not sympathetic and said that if I did not feel able to join the parade, then he would take the parade to me. I was still sitting on the 'thunderbox' when I heard the company being set off and, seconds later, they appeared, marching around the knoll that up to then, had sheltered me. As they passed about 20 yards from me, DBG ordered 'By the right, eyes right' and one hundred pairs of eyes swung sharply in my direction. It is extremely difficult to retain any dignity when one hundred people are laughing at you with your trousers around your ankles! I had enough sense to grab my beret, put it on my head and return the salute. Warrant officers certainly know how to ensure young officers do not get above themselves and the embarrassment was harmless – all part of the learning process.

My application for a regular commission had been endorsed by my commanding officer and a place in the Regiment was available for me. Before I could become a regular officer I had to pay a third visit to the RCB at Westbury. With the support of my Regiment, this time the four-day interview was more of a formality. I duly passed and a full career lay in front of me. I was now committed to spend the major part of my adult working life serving in the Army. I looked forward to it with a mixture of keen anticipation and just a pinch of trepidation about what might lay in store.

On return from Westbury, I was not allowed to stay in Berlin for very long. Young, regular, infantry officers were all sent to complete their infantry training on two courses that followed one directly after the other. Together they lasted several months. For officers trained at Sandhurst, the courses came almost immediately after commissioning. Short service officers did not attend these courses, as their service was too short to justify the very considerable expense, so I had to wait until my regular commission was confirmed before being sent. A satisfactory result from both courses was required (a 'C' pass as a minimum, but a 'B' pass was highly prized).

The first course was held at the School of Musketry, in Hythe, Kent where officers were taught how to instruct on all the weapons in their platoon's inventory; how to conduct range work; and how to plan and conduct live firing exercises. The next course was held at the School of Infantry at Warminster. There student officers were taught about how to command a platoon in much greater detail than was covered at Sandhurst, or at Mons. My two years' experience commanding a platoon proved invaluable. I navigated both courses without too much difficulty, gaining two 'B' passes. That result was a rarity and harbingered a successful career.

I was pleased to get back to Berlin. But a surprise lay ahead of me. This was the battalion's second year in the city and the dreaded platoon battle tests were about to take place. Each of the three battalions held their own rehearsal tests as part of the work up to the actual event. One of our platoons was commanded by a charismatic young officer, Tim Graham Weall. Very unfortunately he had contracted rheumatoid arthritis that effected his joints to the extent that he could not keep up with his platoon. They had come a distant last in the rehearsal tests. In my absence in UK, my original platoon had been assigned to another subaltern. I was a free agent, so I was posted to command Tim's platoon. There were only a couple of weeks before the real event took place and, as ever, training was intense. Suffice to note that my new platoon put in an enormous effort and from being last in the battalion's rehearsal, we came second in the brigade's test and the highest placed of the battalion's platoons. While I was delighted and my professional standing was enhanced, I felt for Tim. He had trained his men well and only his condition prevented him getting the recognition his efforts deserved. His ailment quickly worsened. He had to leave the Army and very sadly died an early death.

The battle test result was a professional highlight for me, but a sporting highlight lay ahead. West Berlin's civilian police force numbered about 15,000. Apart from normal police duties, they were a potent, para-military force, well equipped and able to reinforce the combat power of the allies. They also had a formidable football team. The West Berlin police displayed their para-military combat abilities by mounting an annual tattoo in the impressive Berlin Olympic stadium. The tattoo was on the scale of the Royal Tournament in London. As a warm up to the tattoo, the police had offered to play a football match against a British Combined Forces' team (Army and RAF). It was too good a chance to miss, even though our numbers were about a third of the Germans and they were an established side, while ours would be formed just for the match. I was selected for the British squad.

Kick off was about two hours before the tattoo was due to begin. There were only a few thousand in the stadium at that time, but numbers grew gradually over the first forty-five minutes. The German police were very good and when I took over in goal at half-time we were 2-0 down. As I ran out with the team for the second half, there were possibly only 10,000–15,000 spectators in that vast arena, but it was still impressive. My goal was subject to a constant bombardment and the German crowd made more and more noise as their numbers increased to around 80,000 at the end of the match. We held on for quite some time, but more goals were inevitable and that increased the crowd's good humour. At 4–0 down, a long cross was hit towards my back post. The ball went up and up, for much of the time against a backdrop of cheering people. The noise rose in a crescendo with it. I ran backwards, jumped and made a good catch. The crowd's cheering slumped and then some applause broke out. As I got to my feet to clear the ball I thought 'I've got to have more of that!' I got my wish. No more goals were scored. I had a good game and a wonderful memory.

Having commanded a rifle platoon for three years, I was promoted to lieutenant and the next step was to become a specialist platoon commander. I was to take over the Mortar Platoon in a few months' time. The platoon provided the battalion with its own, integral, indirect fire capability using the 81mm mortar. The weapon is still in service today and has been a remarkably successful weapons system. It has a range of about 4,500m and is capable of firing a high explosive, fragmentation round, two sorts of smoke round and an illumination round. Like all specialist platoons, the Mortar Platoon was manned by experienced soldiers, who had done their time in rifle platoons. With regard to daily routine, they knew their way around without much help or direction and they expected their officer to be every bit as competent as they were.

Shortly after I left Berlin, the battalion would follow me to the UK to begin a posting in Tidworth, but the next stop of very real consequence was Aden.

Berlin's unique position behind the Iron Curtain had made it a fascinating place to serve. I had gained a regular commission and had grown in stature professionally. It had been wonderful socially and I had enjoyed my sport enormously – altogether an experience to be envied.

CHAPTER FIVE
ACTION IN ADEN

In MID-SEPTEMBER 1966, the battalion reformed in Tidworth, a small Hampshire town dwarfed by a huge garrison, sheltering in the lee of the Salisbury Plain training area. It was to be only the briefest of stop-overs. Within days, a few specially selected soldiers flew to Aden, a British protectorate in the Arabian Gulf, to begin training as Arabic speakers and preparations began for the rest of the battalion to take up an anti-terrorist role in the Arabian Gulf before the end of October.

The transformation required was enormous. Berlin was largely about formal guards and duties and a preparedness to confront a threat from the Soviets and East Germans that was implied rather than actual. In Aden the threat was a reality and violent insurrection was rife. Shooting, grenade throwing and rioting were daily occurrences. The battalion had only six weeks to train, before it would be sent to support Aden's small, hard pressed, civilian police force, which had not the numbers, equipment, or expertise to confront a major terrorist campaign. The British government had announced its intention to withdraw from the Protectorate in late 1967 and two Egyptian backed organisations had aspirations to be Aden's government, post the British withdrawal. In the meantime, they pursued a campaign of terror.

In the late 1960s, the Army did not have the custom designed training packages that were developed later to enable units to prepare for specific roles, e.g. peacekeeping in Northern Ireland, or fighting armoured battles in the Gulf Wars. It was very largely left to individual commanding officers to design their own programmes, drawing on the content of the appropriate training pamphlets, which sometimes were not as in date as they should have been. Some of the internal security drills and tactics that we practised were throwbacks to the Malayan Emergency of the 1950s. For example, the anti-riot drill demanded that platoons adopt a box formation and in its centre, apart from the commander, medic and a radio operator, there were stretcher bearers, banner men, who, at an appropriate moment, would unfurl a banner that read something along the lines of '*Disperse or we fire*', and a bugler to draw rioters' attention to the banner. The platoon commander was also required to read the Riot Act before taking any action. Actually, the 'Malayan Emergency' might be a bit generous. That tactic was the stuff of the Indian Mutiny of 1857 and absolutely no good at all when rioters were well armed and had grenades to throw at you. It was used once before we knew better ... very much better.

Despite these limitations, an intense and relevant training programme was followed. We concentrated on fitness, snap shooting to engage fleeting targets and orientation briefings that covered everything from personal care in hot climates, to the political situation and the adversaries we would face. We drew heavily on advice from the 1st Battalion, Somerset and Cornwall Light Infantry, the unit we were to relieve.

Every soldier had a series of injections against various diseases and had to pass a detailed medical examination before being permitted to deploy. We also had to be kitted out with hot weather clothing – 'KD' (khaki drill). Buff coloured, it consisted of aertex-type shirts, long trousers, a soft hat (always called 'hats ridiculous' for reasons lost in the mists of time), cotton underwear, shorts, that came with dark blue hose tops (long socks without feet) and purple garter flashes (small pieces of material that protrude from under the folded top of the hose tops – for decoration and to help keep the hose tops up – another throw-back to colonial soldiering). The shorts were enormous to the extent that small soldiers were in danger of taking off in anything above a stiff breeze.

The rapid turn around and deployment was difficult enough for the men, but it was doubly difficult for the battalion's wives and children. The overseas allowance was lost on return to England, so money was tighter. In Berlin families had become used to well-maintained quarters, largely paid for by the West German funded Berlin budget. In contrast, Tidworth was an old garrison. The barracks dated back to the Crimean War and many of the married quarters were in dire need of modernisation. Tidworth was tired and tiny, with only a handful of shops and had few civilian amenities to add to those the military provided. It was also comparatively isolated – Andover, the nearest, reasonably sized town, was some 10 miles away. Most of the children had to be settled into new schools and for many, it was not an easy process. The wives had six weeks of contending with all of that before their husbands were sent thousands of miles away, into an active theatre and their constant worry became 'Will he come home?' In Aden, telephone communication was not available to anyone and the only means of receiving news was by letter, normally 'bluies' (flimsy, single sheets of paper, with gummed edges that folded over to make an airmail letter). Wives had to contend with a lot. Within the Army community, it is not only the soldiers who deserve medals.

Just before I left for Aden, Sue and I had arranged to meet. All the time I was in Berlin, we had kept in touch by occasional letter. Sue was now a qualified librarian and lived and worked in Alcester, in Warwickshire. She had developed doubts about her fiancé and the engagement had not worked out. It was called off and I was back on the scene. We picked up where we had left off and, for my part, I certainly appreciated the person I had been missing for nearly two years. I was sure Aden would be exciting, but other than my immediate family, it was good to know that I had someone very special waiting at home.

Early on the morning of 23 October 1966, civilian coaches drew onto the battalion's parade square and, along with the rest of my company, I climbed aboard. As we drove along the A303 towards RAF Brize Norton, I looked out on a superb English autumn morning. The sky was Wedgewood blue. There was a hoar frost on the grass, which glistened in crystal clear sunlight. The trees had retained most of their leaves, which varied from green, through gold and amber, to a deep russet. It was magnificent. I tried not to be maudlin, but I could not help wondering if I would see a scene like that, or even England, again. I forced myself to get a grip and look forward to what lay ahead.

The military have a saying, 'Time to spare, go by air'! At Brize Norton the RAF movements' staff were their usual officious, unhelpful selves and we spent hours hanging around before our flight took off. We were to travel by a *Britannia* turbo prop airliner. The journey would be very long, very tiresome and involved at least one, if not two, re-fuelling stops. At last, with the evening sky darkening, we began our descent into RAF Khormaksa, in Aden. In 1966 it was the RAF's biggest base. As the *Britannia*'s door was opened a rush of extremely hot, horridly humid air flooded in. In seconds everyone was sweating and we had yet to set foot outside. It was an immediate foretaste of the climatic conditions that awaited us.

The flight was met by our commanding officer (CO), Lieutenant Colonel Peter Leng, and the Regimental Sergeant Major (RSM). They were not their normal ebullient selves and the CO asked that all the soldiers on the flight be assembled.

The CO said while it was the last thing he wanted to do on our arrival, he needed to break some quite dreadful news. He informed us that earlier that evening, there had been a patrol clash between a Special Air Service patrol and one of our own. There had been an exchange of fire and Corporal Watkins, who was commanding a patrol from the Recce Platoon, had been killed. Obviously, the clash should not have happened and the CO explained that both patrols were aware of the other's presence. A block of housing had been left between their respective patrol areas and they had been warned not to enter it, but for whatever reason, that warning had been ignored by both and the clash resulted. We listened to the CO in a profound, shocked silence. We had had a good man killed and we, the new arrivals, had not even reached our camp.

Radfan Camp was a tented camp, adjacent to Khormaksa airfield and only about 400m from the sea. It was built in desert, directly onto the sand. There were very few brick or metal buildings: cookhouse, NAAFI, officers' and sergeants' messes, battalion headquarters and the armouries. The wash blocks were wooden framed with 'wriggly tin' walls, with gaps top and bottom to allow air circulation. Washing water was not heated: all that was required was for the water tanks to be mounted on the roofs and the sun did the rest.

The camp had rows and rows of brown marquees erected on concrete bases (the rows were called 'lines' from the days of the Army in India, when tents were erected in immaculately straight 'lines'). These provided sleeping accommodation for everyone – officers, NCOs and soldiers alike. A marquee provided eight to ten bed spaces for soldiers and JNCOs. Officers and SNCOs were given more space. 'Air conditioning' was obtained by taking the sides off the marquees, to allow a free flow of hot air that was further circulated by two, large, ceiling mounted fans. To provide protection from terrorist home-made mortars, each bed space had waist high, breeze block walls built around three of its sides.

Radfan Camp housed two battalions. We shared it with the Cameron Highlanders. They operated in the main town of Aden, which was several miles south of us, while we were tasked with maintaining security in Sheikh Othman and its satellite, the residential area of Al Mansoura. Both lay about 4 miles to the north and together with an expanse of desert and a large area of salt pans, they formed 'Area North'. Immediately, north of Sheikh Othman was the boundary with the Sultanate of Lahej, one of the original nine cantons that had signed protectorate agreements with the British in the nineteenth century. The battalion did not operate in Lahej.

Two quasi-political, terrorist groups confronted us – the Federation for the Liberation of South Yemen (FLOSY) and the National Liberation Front (NLF). They had two identical objectives. First, to claim that they had been largely instrumental in forcing a British withdrawal, thus their attempts to kill as many of us as possible; and, second, to become the dominant political power and the faction most able to provide a future government. Open conflict existed between them. Like terrorist organisations the world over, they preyed on the civilian population, most of whom simply wanted to get on with their own lives and were heartily sick of the violence that surrounded them. However, the effective intimidation of normal civilians meant that they were loath to have anything to do with us, thus, the flow of even low-level intelligence was limited and high-level intelligence was virtually non-existent. At the beginning of their campaign, the terrorist groups had targeted the Aden police forces' Special Branch operatives, murdering many of them. That largely nullified their sources and meant we were operating almost blind, except on the odd occasion when we could generate our own intelligence from a telling arrest, or the discovery of an arms cache from which evidence could be extracted.

The day after arriving in Aden I was walking through camp when I met the CO.

'Have you been into Sheikh Othman yet?' he asked.

I said that my first patrol was not scheduled until the following day.

'Get your rifle and equipment and come with me. I'll give you a look around.'

I ran to do as I had been told and joined the CO and his crew at the loading bay. We put loaded magazines on our weapons and got into the two Land Rovers that were for the CO and his escort. The escort was commanded by the RSM. There is always a close relationship between the CO and RSM, but on active tours they become inseparable.

We drove to Sheikh Othman and headed for all the major landmarks that I had heard and read about – grenade corner, the main mosque and square, the principal through routes, the post office and police station. We drove through Al Qahira, a suburb of Sheikh Othman, consisting of a grid of unpaved streets, lined by scruffy, low rise buildings. It was a major centre of trouble. We continued to the cattle market and the Al Mansoura roundabout. The CO had stopped at the main square, allowing me to get out and to move amongst the locals as they went about their business.

The sights, sounds and smells were something that no amount of briefing could convey accurately. Men were dressed in flowing robes, or *futahs* (a wrap-around length of soft cotton to make a loose form of skirt) and had various styles of turban on their heads; women wore *hijabs*; wandering cows, and camel and donkey carts weaved through chaotic motor traffic, barely giving way to massively overloaded, highly decorated lorries and buses. Apart from motor vehicles, there was not a lot of colour; whitewashed buildings and dun-coloured sand predominated, punctuated by advertising signs and the black and brown robes of adult women. There was the strong smell of cooking and spices and the much less pleasant odour of rotting rubbish that littered all the streets. The CO had thoughtfully given me a gentle introduction to the environment that I would get to know so well in coming months. I felt much better prepared to lead my platoon when it conducted its first patrols the following day.

The duties we were to perform were varied. Companies and their platoons rotated through a series of tasks:

- The inevitable guards for Radfan Camp itself; for Mansoura Detention Centre that housed convicted and detained terrorists; and sundry installations like the British Forces Broadcast Services' radio station, because anything British was a target.
- Checkpoint duty. Within our area of operation there were four large, vehicle checkpoints, there to monitor and search traffic entering and leaving Aden. The two larger ones took a platoon to staff them. A third platoon was divided between the smaller ones. Checkpoint duty usually lasted for twenty-four hours, but could be extended as necessary during periods of heightened trouble. Checkpoints were high visibility and, at least initially, we were required to wear starched shirts, shorts and our woolly hose tops while manning them. Apart from the command tent, search area and guard towers, checkpoints also had an administrative area with a kitchen tent, dining and sleeping marquees, toilets and a rudimentary shower facility.
- Patrolling, which took various forms. Most commonly, platoons were

tasked to patrol on foot within the urban areas of Sheikh Othman and Al Mansoura. During periods when rioting was either threatened or taking place, foot patrols were supported by armoured cars from The Queen's Dragoon Guards, when terrorist gunmen and grenade throwers provided the most common threat. Land Rover patrols were used to police the salt pans and desert areas, reinforced occasionally by troop carrying helicopters from the RAF.

- Route clearance and ambushes. A favourite terrorist tactic was the roadside bomb. Some were anti-tank mines, often dating from World War II, set off by vehicles driving over them – the vehicles were not always ours. Other bombs were electrically detonated having had a command wire laid to them from a remote firing point. Depending on the type of device used, both vehicles and personnel could be targeted. To combat the roadside bomb threat, night time ambushes were set overlooking stretches of road that terrorists would be most likely to target. Early every morning, route clearance patrols were conducted. Our principal clearance aides were rudimentary: the eyeball, to look for traces of ground disturbance and a very heavy metal hook on the end of a pole. The hook was dragged by a soldier holding the pole, with one such soldier deployed on either side of a road, so that the hook sank into the sand and, with luck, would snag any command wire that had been laid.

- Cordons and searches. Despite the paucity of intelligence, some was gleaned and that often led to searches of both commercial and residential property. While the searches were in progress, the area had to be secured by a cordon. These were manpower intensive operations that when possible, were conducted at night.

My platoon, 15 Platoon, settled into those various duties in the weeks that followed. It was a lively time for most, but not for us. Weeks turned into months and still 15 Platoon had no cause to fire a shot in anger. Terrorist incidents, of one kind or another, occurred daily. While every other platoon had been engaged, the terrorists seemed to be avoiding us. Of course, we boasted that this was down to our professionalism, but the truth was that we were all a bit on edge wondering when our baptism of fire would take place.

Just after midday on the 5 December, with many of the soldiers returning from lunch, there was the repeated 'crump, crump' of mortars being fired at Radfan Camp. Seconds later, rounds started to explode in the camp. The sergeants' mess took the brunt of the attack and its mess members suffered seven slight casualties. The firing point was some way outside the camp. When it was located it became clear that thirty home-made, pipe mortars had been set up with a rudimentary timing device used to fire them. Of the thirty mortars, only twelve fired; three dropped short of the camp; four failed to explode on impact; and only five did the job for which they were intended. It was fairly typical of the terrorists' self-made weaponry, but they only had to get things right once to cause mayhem and distress.

Checkpoints provided tempting targets, although they were not particularly dangerous places to be. It was during one checkpoint duty that 15 Platoon got its first taste of action. The checkpoint in question was at a key intersection with a road leading south from Al Mansoura that branched to Aden town in one direction and to the satellite settlement of Little Aden in the other. Around midday an enormous explosion was heard, followed by a dust cloud billowing over Al Mansoura. The radio net burst into life and it was immediately clear that 15 Platoon was best placed to react. The off duty watch crammed onto two Land Rovers and in about two to three minutes we were on the scene. It was carnage. The fronts of two buildings had been blown apart, together with the perimeter walls to the properties. For some considerable distance, windows had been blown in and shattered glass littered the road. Body parts were strewn everywhere. Dust covered, bleeding, bewildered people lay, or wandered around and some uninjured, traumatised women had begun to keen. Would be helpers were already doing what they could to provide first aid and comfort.

Our first task was to preserve the scene for the civilian Scenes of Crime Officer in order that he would be able to conduct a forensic examination. That entailed throwing a cordon around the immediate area of the explosion, keeping those from outside from entering, while containing the unfortunate survivors inside, to receive skilled first aid, initially from our medic, until the civilian emergency services arrived, and to hold them there for the Aden police to record witnesses and take statements in their own time. We got all that done fairly quickly, aided by reinforcements from the Mansoura Detention Centre guard and elements of the company on patrol duty, who arrived shortly after us.

I had radioed an initial contact report as soon as we had got to the scene, but now I had time to send a more detailed, considered sitrep (situation report). I did that and then went around the cordon to check on my men to make sure that they had taken up the best possible positions and they were clear on what was happening. I got to one newly arrived soldier who was serving his first day of active duty. I was concerned that the awfulness that confronted him might just be too much for such an inexperienced, teenage soldier.

'Did you report three dead, sir?'

'Yes,' I said.

'Well, I think there are seven legs, or at least bits of them.'

His presence of mind was remarkable and he was right. One sad remnant of a human being had been thrown into the rubble just beyond his fire position and I had not seen it until he pointed it out. It was a truly remarkable, cool, collected reaction for one so young. Soldiers never ceased to amaze.

A week or so later, 15 Platoon were manning another checkpoint at the border with Lahej. Checkpoint guards did not just sit tight inside the checkpoint's perimeter, but sent clearance patrols out into the surrounding area to deter terrorists from setting up attacks. Most checkpoints were located in wide, open areas, but this one was sandwiched between a college on one side of the main road and a sprawling area of kutcha huts on the other. These ramshackle huts, built directly onto the sand, were made of anything that came to hand, from wrought iron and plastic sheets, to bits of wood and cardboard. Most were the size of a modest, family tent. It was difficult to ascertain what was going on only yards into this haphazard conglomeration of constructions, making them ideal cover for terrorists.

In the late afternoon we heard the now familiar 'crump, crump' and took cover. We waited for the best part of a minute before we were certain that no explosion would take place. The sentry in the tower overlooking the checkpoint confirmed the direction from which the crumps had come. My excellent platoon sergeant, Sergeant Mick Dear, was on duty at the time and immediately took some men into the kutcha hut area to investigate. I suspended vehicle and pedestrian movement at the checkpoint, then went with three or four soldiers to see if I could find any evidence of what had been fired at us. In the coils of barbed wire that formed the perimeter fence, we found two energa grenades (British Army munitions), one badly dented, the other was pristine.

Energa grenades were anti-tank projectiles, designed to be fired from the end of a rifle, but they required a special cartridge to do so. Without that cartridge, the terrorists had to improvise and had used a smaller version of the pipe mortar launcher to propel the grenade. The grenade would only detonate if it hit a hard surface, like a tank's armour, but as about eighty per cent of the checkpoint was sand, they had failed to explode. The terrorists' ingenuity had to be admired, but fortunately they lacked an understanding of the grenade's fuse mechanism, so their attack was doomed to failure. It was hardly an incident to make the headlines, but at least 15 Platoon now felt worthy of being targeted.

While the soldiering was serious, Aden was not all work and no play. It had been a key naval base for a very long time, strategically important as it guarded the entrance to the Red Sea and therefore, it is the eastern gateway to the Suez Canal. The naval installation at the port was extensive, with many sporting and recreational activities, including secure beaches. We made as much use as possible of the beaches to relax and, importantly, to meet some of the very few young women that were in Aden at the time, usually the daughters of officers serving two-year tours in staff roles in headquarters. Just being able to talk to a woman brought a pleasant degree of normality to our lives.

Aden was a free port and it was still an important calling point for passenger liners. The shopping areas, one adjacent to the port and another close to the RAF base, managed to continue a vibrant trade in custom free goods during the insurrection, as did the few good restaurants. With a considerable amount of care, occasionally we could go shopping and have a meal out.

Another welcome departure from normal duties came when I was nominated to attend a forward air controllers' (FAC) course, held locally. RAF Khormaksa was home to two Hunter jet fighter squadrons, whose principal role was ground attack in support of Army units. They were no use at all in urban areas, but up-country Aden also had its terrorists and the vast majority of the land there was mountainous desert, ideal for the Hunters to operate.

Controlling aircraft that travel at several hundreds of knots per hour, was a tricky business. When a strike was ordered, aircraft were held at an initiation point, probably about 20 miles from the target, in order to preserve surprise and to keep the aircraft safe while the pilots were briefed by radio by the FAC. The FAC gave the pilot a heading and a time to 'run in' to the target at a set speed. The 'run in' was done at a maximum of 150 feet and when the 'run in' time had elapsed, the pilot pulled up sharply to gain a height of about 2,000 feet. Although the FAC knew the direction from which the fighter was coming, the pull up might be the first time he saw the aircraft. There were only a few seconds for the FAC to describe the target in detail and for the pilot to identify it, before the aircraft executed a dive and delivered its weaponry. Today targets are illuminated by lasers, but then fast, accurate, descriptive talking was required.

To give FACs a pilot's eye view, we were flown in a two-seat version of the Hunter. It was an extraordinary experience. My pilot flew me up country from Khormaksa and then the fun began. We did a 'run in', pull up, wing over and dive onto a dummy target. 'We' did not include my stomach, which was left somewhere behind after the aircraft pulled up. Fortunately, my breakfast stayed with me and we did two more attacks before the pilot took pity on me. He climbed to a considerable height. His darkened, bone dome helmet turned towards me.

'Would you like a go?'

I could hardly believe it. 'Are you kidding? Yes, please.'

He explained that I was not to touch the rudders, but would have charge of the control stick. If he said 'I have control', I was to release it immediately. I was told to put my hand on the control stick, but not to grip it until I was told 'You have control'. When that came, I confirmed 'I have control' and found myself flying an aerial thoroughbred more or less straight and level. It was exhilarating. 'You seem to have got that. Now try a gentle turn to the left.' I moved the stick and suddenly I was flung back hard into my seat as I mistakenly pulled several 'G'. Instantly the pilot said 'I have control' and normality was returned as quickly as I had made it disappear. After a bit more tuition I was able to turn the aircraft without my face changing shape and it was time to go home. The aircraft dropped down to something like 50 feet. The ground rushed past at an alarming rate and we were flying down deep *wadis* (valleys), with cliff faces either side of us, banking left and right as we followed dried up water courses. Alton Towers and Blackpool do not have rides that are in any way comparable. Amazing flying and a memory of a lifetime.

Just before Christmas 1966, my platoon and I got a bonus. We were tasked to provide the camp guard at Mukeiras, which was situated only a couple of miles from the border with the Yemen Arab Republic, in high, mountainous desert, about 100 miles from Aden. The Arab settlement there was not much more than a large village, built almost entirely of mud. It was the seat of a local, feudal *naib* (lord) who controlled absolutely everything within his area. The way of life up country had changed little since the dark ages. All adult men had a curved dagger in their waistbands and carried a rifle of some description, ranging from ancient muzzle loaders to modern Kalashnikovs. The town was built at one end of a wide valley on which an airstrip had been constructed. Next to the airstrip, and a mile from the Arab settlement, was an RAF radar, there to give early warning of any aggressive flights from the Yemen. The Army camp was built around the radar and its troops were tasked to defend it.

A platoon was required to guard the camp, but because it was so isolated from any other British unit, its firepower was significantly reinforced by two 105mm howitzers from the Royal Artillery and we also took a section of our own 81mm mortars with us. We were a significant force.

We flew to Mukeiras in a *Dakota DC5* of Aden Airways, an off-shoot of BOAC (British Overseas Airways Corporation, now British Airways). We landed and marched the few hundred yards from the aircraft pan to the camp. A handover from the outgoing platoon was completed quickly and they were away within an hour of our arrival. Only then did I meet the camp commandant, a major from another of the Aden based battalions. His battalion was responsible for the administration of the camp and it also provided the only communications link to Aden. The distance was so great that only a Morse link, using high frequency (HF) radio, would work and then only in daytime as night time atmospherics make a HF signal unreadable. I had no direct communication with my own battalion.

I thought it a little strange that a major should be the camp commandant of such a small installation. When we did meet, I thought it even stranger. He was an ex-heavyweight boxer and my first impression was that he might be punch drunk. We were to be at Mukeiras for several weeks and as there was only him, an artillery officer and me in the mess, there would be no avoiding one another.

Our first duty, after taking over the guard, was to settle into our accommodation – marquees, exactly the same as we had left in Radfan Camp. This was done quickly, then a tour of the camp was organised because, even though it was only about 100m to 150m across, we had to know exactly where everything was and be able to find our way in the dark – no outside lights were allowed. Every evening, as night fell, and again at dawn, the entire camp 'stood to'. 'Stand to' required everyone to be armed, in fighting order, in dug-in fire positions, observing their arcs of fire. The artillery and mortars registered (checked the data) and fired on some of their defensive fire tasks. Artillery shells and mortar rounds, exploding on areas that terrorists might well use to attack the camp, were a very effective deterrent. This was serious bandit country.

I took up my position in the command bunker. About ten minutes into the 'stand to' period a field telephone rang and one of my corporals said that he wanted me in his bunker because the camp commandant was crawling around inside the several layers of barbed wire fencing that formed the camp's perimeter. I found it difficult to believe, but the corporal was insistent. I left the command bunker to investigate.

The bunkers provided sentries with protected fire positions. They were dug into the rocky ground and to protect the occupants from mortars, they had 50mm of rocky spoil as overhead cover. They were substantial constructions that could take three or four men. I shouldered my way in. Off to one side, about 30m away, I could see the figure of a man crawling towards us. That was an incredibly dangerous thing to do. Quite apart from risking the camp guard firing on him, he also risked setting off the claymore anti-personnel mines and magnesium trip-flares that were positioned amongst the fencing. We waited.

The figure got to within 1m of the bunker and then said, 'You're all asleep. Get your platoon commander up here!'

'I'm here, sir,' I said. 'What are you doing? My chaps have been watching you for the last ten minutes.'

'I'll see you in the mess,' was the only response I got.

After 'stand to' was over and the normal night manning of sentry positions began, I went to the mess. The camp commandant claimed that he went through the performance that we had just witnessed, every time a platoon took over its guard duty. I was more than ever convinced that the man was ill, but could do nothing about it. The HF link was manned by his signallers and terminated with his battalion in Aden. I could hardly ask the signallers to send a message saying that their camp commandant was ill to the point of being deranged. I would just have to work through it until I could contact my battalion directly.

Despite that, the time at Mukeiras was wonderfully interesting. Every day we patrolled to local villages 'to show the flag' and let their inhabitants know that we were around. We even played an impromptu game of football, on a rock-strewn pitch, against locals from one village. Our weapons were carefully guarded by non-participating colleagues on the side lines and the most interested spectators were our two machine gunners providing over watch of the entire proceeding.

We had two medics (corporals) from the Royal Army Medical Corps (RAMC) with us. They always accompanied us on local patrols and held pop-up clinics in the villages that we visited. Their efforts were particularly welcomed by the locals because even rudimentary medications seemed to have remarkable effects on people who were otherwise without medicine. The senior corporal warned me that in one village they had been treating a woman who had tripped and poured hot fat over her breast and stomach. The flow of scolding liquid had only been arrested by her waist belt. It was remarkable that the medics had been asked to treat a woman, particularly so intimately. They had no really suitable antibiotics and were making do with shell dressings, morphine and cicatrine powder, something used to dress gunshot wounds. The senior corporal made clear to me that the normal thirty or forty minutes that we stayed in villages, would not be enough time for the treatment of the woman's wound. I adjusted my patrol plan accordingly.

On arrival at the village, the patrol deployed into defensive fire positions. I spoke to the headman using sign language and the little Arabic that I had picked up on checkpoints. The medics set up their clinic, which consisted of two backpacks. When the injured woman's adult son appeared, the corporals made to go with him, together with an escort, but the son noticed that one of the corporals was not wearing his maroon, RAMC lanyard. For some reason, villagers associated a lanyard with being a doctor. Royal Anglian soldiers do not wear lanyards, but their officers and SNCOs do. Mine was Pompadour purple and the colour was not that far removed from the RAMC's maroon. The second corporal was not allowed into the house, but because I had a lanyard, I had to take his place.

The senior medic said that it would not be a pleasant experience. He gave me a face mask to wear, with a couple of drops of something that smelled like pear drops on it. It was powerful, but it was completely overcome by the smell of gas gangrene when we got to the lady, lying on mats, on the flat roof of the family's home. The medic gave her quite a lot of morphine and while we waited for it to take effect, we began cleaning the wound. It was a vast area, hugely discoloured, suppurating pus and, in that fly-blown environment, there were maggots. I was constantly at the point of being sick. Eventually the wound was cleaned and the morphine had done its stuff. Now the corporal took a scalpel and began removing bits of totally rotten flesh. When he had done that the entire wound was treated with cicatrine and bound with a couple of fresh field dressings. Had she been a western woman, unused to severe physical hardship, it is likely that an injury of that severity would have killed her. The medic had done all he could and his efforts were remarkable. We left the lady in the care of her family and re-joined the patrol. I have never been so pleased to get back into fresh air in my life. I recommended the corporal for an award, but after my return to Aden, I never saw him again, or learned the fate of the lady.

Two or three times a week, we mounted ambushes on mountain trails that we thought terrorists might use. These were not the relatively low key, route ambushes that we employed in Aden. Here we were after gun runners and terrorist groups who normally took shelter over the border in Yemen. We took as much firepower with us as we could, so apart from our rifles and machine guns, we had claymore mines and grenade necklaces (hand grenades, linked together with a long length of special fuse, to explode simultaneously when fired). In addition, while we were in ambush, the artillery and mortar men manned their weapons, which were laid on targets immediately adjacent to the ambush site. From a professional point of view, it was regrettable that all the planning and rehearsals that went into those ambushes, came to nothing, but it was all valuable experience for us. The biggest downside was the bitter cold to be endured in the desert at night, while lying perfectly still. Conversely, the clear, glittering, star-studded, night sky was magical.

Towards the end of my platoon's duty at Mukeiras, I was informed that my commanding officer was to fly in to visit us. I welcomed this first contact with my battalion since leaving Aden, but it brought its own problem. While we had been up country, the commanding officer had changed. It was no longer Lieutenant Colonel Peter Leng. Instead I was to have my first ever meeting with Lieutenant Colonel John Dymoke. I met him at the airstrip and asked if he minded walking with me to the camp. I explained that I wanted to speak to him privately and as a matter of urgency. This was not a request he was expecting on first acquaintance, but apart from raised eyebrows, he agreed. I explained the position regarding the camp commandant. My new CO took note, but understandably reserved judgement until he had met the officer and spoken to my men. The visit was a short one. The CO spoke to the camp commandant to form his own opinion and then spoke to Sergeant Dear and my other NCOs. They all supported me and some had their own stories to tell. A couple of days later we were on a plane back to Aden with our job done. It transpired that the camp commandant was ill and he was replaced very shortly afterwards.

While my platoon had been away, the security situation in Aden had worsened as independence from Britain drew closer and FLOSY and the NLF redoubled their efforts to become the dominant power. Terrorists promoted strikes; small scale rioting became increasingly common; and there was an intensification in grenade and shooting attacks. Throughout March it became clear that the terrorists were attempting to make Sheikh Othman a no-go area for the British. Because of the intensity of operations, the battalion received reinforcement from a company the King's Own Border Regiment, normally based in Bahrain, and occasionally, we had a further company attached from 45 Commando Royal Marines, based in Little Aden. The patrol effort was intense and like everyone else, 15 Platoon was in the thick of it.

A UN mission to Aden was announced to arrive on 1 April, but on that day the heavens opened and Aden suffered its first floods in living memory. The weather only delayed the mission and the strike and insurrection that the terrorists had arranged to coincide with it, by twenty-four hours. On the 3 April co-ordinated terrorist attacks increased to a level we had not experienced before. They were led by Egyptian trained commandos and street fighting, rather than peacekeeping, was commonplace over the ensuing days.

On 6 April my company was deployed to an area south of Sheikh Othman and was held outside the town. A *Saladin* armoured car had been blown up by a mine concealed in rubbish that formed part of one of several barricades, hastily constructed across some of the roads in the town. My company commander, Major Roy Jackson, called for me. He wanted my platoon to clear as many of the barricades as possible and showed me the route he wanted me to take. We set off in two armoured 4-ton trucks, de-bussed just short of the town and then patrolled on foot, making for the main square. There were few people about and virtually no traffic. We cleared a couple of barricades, being extremely careful in case booby traps and mines had been laid. We then moved on to the narrower streets of Al Qahira to continue barricade clearance. Local people left the streets when they saw us coming. That was not a good sign as it left the way clear for terrorists to attack with less chance of injuring the local populace. In the event, nothing happened and we returned to our company's holding area.

An hour or so later, I was again called to my company commander. Major Jackson wanted the patrol repeated, but the route was to be exactly that which we had taken previously. I had a high regard for my company commander, but felt that this would establish a pattern and it was a fundamental mistake. We had a sharp altercation, but I was not in a position to contradict my senior officer.

Sheikh Othman was even quieter than we had found it during our previous patrol. The barricades that we had cleared had not been re-established in the mosque and square area. The atmosphere was tense and we worked hard dodging from one bit of cover to the next as we advanced along the route. In Al Qahira, I split the platoon to advance in parallel along two streets to make it more difficult for terrorists to observe us easily. Suddenly, on my street, there was nobody but us. About 100m ahead of us a gunman appeared from around a corner, fired a prolonged burst at us and ducked back into the cover of a building. Training kicked in; soldiers took cover instantly and looked to return fire. Then the gunman made a mistake – he reappeared and fired a second burst. My point section, led by Corporal Sean Sweeney, was the first to react. Two shots were got away and the gunman was hit. He was still 70m or 80m from my nearest man and before we could get to him, he had been dragged into a building next to where he had fallen. It was a small mosque, so we could not enter and people were coming out of their houses to shout and throw rocks at us. The locals' presence on the street indicated that the attack was over. There was nothing for it but to move on. We completed the patrol without further incident and returned to snatch some rest in the holding area.

A short time later, a fellow subaltern commanding 13 Platoon, Lieutenant Brian Harrington-Spier, was ordered to take on the same route clearance task. I cannot be certain that he was told to take exactly the same route that had been followed by my platoon previously, but it was very similar. Brian split his platoon as I had done and as his half entered the main mosque area, a couple of grenades were thrown at it, causing one slight casualty. Further into the square, his half of the platoon was engaged by heavy fire from multiple positions. Brian ordered a withdrawal, under fire, to a pre-arranged rendezvous point. When the entire platoon got to the RV they discovered that two soldiers were missing, still somewhere in the square. Brian returned, accompanied by a *Saladin* armoured car and a *Saracen* armoured personnel carrier. Suffice to record that Brian recovered his men. For his bravery he was awarded a Member of the British Empire medal. Corporal Valentine, a section commander, who accompanied him throughout, received a British Empire Medal. Both medals had crossed oak leaves on their ribbons, which unlike the vast majority of MBEs and BEMs, means that they were, very deservedly, awarded for gallantry.

Terrorist activity did not diminish throughout the day. The main mosque and square were where most of the shooting took place. In the mid-afternoon my platoon was ordered to occupy the flat roofs of two substantial houses at a road junction known as 'The Obelisk'. The Obelisk itself was about 600m south of the square and the view into Sheikh Othman from there was excellent. My orders were to neutralise terrorist gunmen firing from the rooftops of the tall buildings on the square's southern boundary.

My platoon did not have to wait long. We had been observed taking up our positions and an over-zealous gunman could not resist firing a few ineffectual bursts in our direction. By way of explanation, an average rifleman can confidently expect to hit targets out to 300m and beyond that, out to 600m, effective harassing fire can be produced. It follows that the platoon and certainly the shot-happy zealot, were operating at the limits of a rifleman's range. However, my machine gunners were in their element, benefitting from a heavier weapon and the stability provided by the weapon's bi-pod legs.

For most of the time, the terrorist who had fired at us, was hidden from view, having taken cover behind a low wall that surrounded the top of his roof. From time to time he would pop up to fire badly aimed bursts and disappear again. He became careless. Through my binoculars I could just see the top of his head. I had a loud hailer with me and I gave a fire control order to my platoon. On my order, the entire platoon would open fire simultaneously. The riflemen were to fire very deliberate shots at the precise point of the rooftop wall that I had indicated to them. The machine gunners would add short, controlled bursts. I could see my soldiers settle behind their weapons. 'Fire' – and a platoon volley rang out. 'Stop'. Through my binoculars I could see chunks of breeze block had been blasted off the wall, but there was no sign of the gunman. Suddenly, he was in view, climbing over the wall onto a metal ladder that ran down the side of the building. I gave another fire control order over the loud hailer. The terrorist was halfway down the ladder when I ordered 'Fire' and he died. I was relieved to remove the threat he posed, but felt nothing for the individual.

For some, using a platoon to engage a single gunman may appear disproportionate. However, in virtually all other circumstances, the balance of advantage lay heavily with the terrorists. They could choose the place and time of their attacks. They could plan their escape routes and post lookouts to give them warning of our movements. Virtually all of their shooting attacks were fleeting. Unless the terrorists made an error, as had been the case earlier that day, or we had an enormous slice of good fortune, we were very unlikely to have the opportunity to engage them successfully. When a chance did come, it had to be seized, or another terrorist was likely to live to perpetrate more pre-planned acts of unnecessary, unjustified violence that could lead to innocent deaths and injuries.

Only a few minutes later we were fired on again, this time from three different positions lower down in the buildings that were facing us. The gunmen had abandoned the rooftops, but despite their commando billing, they were poorly trained, in that they were firing from the frames of windows, instead of further back in the rooms they were occupying. Their forward positions made them easy to spot. Although they had enough sense to change rooms from time to time, we took them out, one by one, and in the course of the next hour, we accounted for at least another four. There was no pride or pleasure taken in that. We were simply doing a dirty job efficiently.

In the early hours of the next morning we searched the blocks of houses from which the gunmen had operated. We found lots of spent cases, bullet damage and dried blood, but the normal occupants had long gone and what was left of the commandos had disappeared too. They had failed spectacularly. In the first week of April the battalion recorded 160 attacks against it. The Quartermaster estimated that during those same few days, the battalion had fired more rounds than any other since the Korean War. We had definitely had a lively time.

The remaining month of the battalion's tour in Aden passed at not quite the same intensity. Major incidents flared from time to time and routine patrolling, route clearances and ambushes continued to be conducted. Checkpoints still had to be manned and during my last checkpoint duty I was forcefully reminded of home. I had drawn the 'dead man's watch' covering the small hours of the morning. There was nothing happening, until we heard the high-pitched whine of Land Rover tyres on tarmac. The re-supply run was arriving, bringing highly prized mail with it. I was given a particularly fat envelope and as I took hold of it, a series of photographs fell out. My signaller scooped them up before I could move, gave a whoop and handed them around. They were of Sue. Leg pulling became the instant order of the day, but the consensus was that here was a lady for serious consideration and that I had better get a move on when I got home. Good advice.

In May, the 1st Battalion, the Parachute Regiment arrived to relieve us and to see out the insurgency until the British handover of power on 30 November 1967. Most of us left the Protectorate without a backward glance. Over the course of the seven months that the battalion was deployed in Aden, more than 450 terrorist incidents were directed against us, resulting in 174 casualties requiring hospital treatment, but thankfully all of them made good recoveries. We had suffered only three deaths and two of those came from a natural cause and a traffic accident. Despite the appalling start, it had been an unbelievably lucky tour of duty.

Our time in Aden had been hard and demanding, both physically and mentally, but it had been rewarding. Many of us had arrived as boys and had become men. All of us were better soldiers for the experience. Lifting off from Khormaksa we knew that the violence, mayhem and oppressive heat were behind us. Now there was leave and family to look forward to and, for me, Sue was foremost in my mind.

CHAPTER SIX

TIDWORTH, ALDERSHOT
AND PENETRATION

ON ARRIVAL BACK in the UK, I made a beeline for London. Sue had moved there and shared a Thames side flat in Chiswick with Nicola, a mutual school friend. Sue now had a job as a qualified librarian in central London. I ran up the stairs to her flat and rang the bell. I heard footsteps approach from the other side of the door. As the lock turned, a draught caught a large wind chime that hung just inches behind my head. It made a loud clang. Sue opened the door to an empty space. I was crouching, two or three steps back down the stairs. Loud, unexpected noises would continue to alarm all of us who had returned from the mayhem of Aden, for quite some time to come. We did not recognise the tension in one another, but to those who had not shared our experience, it was glaringly obvious.

Notwithstanding the unusual start to our reconciliation, we were delighted to see one another, but Sue had an ulterior motive. I had turned up on the day that she and Nicola had arranged to move to another, more spacious flat in Chiswick. The river side flat was on the first floor; the new flat was on the fourth – a fit, young soldier would be extremely useful to help transport their goods and chattels. When the move was over, Sue and I had time to realise just what we meant to each other and we settled into a deep, profound relationship that burgeoned. We still nurture it. It remains fresh, growing day by day.

When the battalion re-formed at Tidworth following leave, its attention was focused on its new role in 24 Airmobile Brigade. As the name suggests, this brigade was designed to deploy rapidly anywhere that a British military intervention was deemed necessary. A brigade is a combined arms formation that includes armour, infantry, artillery, engineers and logistical elements and is about 5,000 strong. 24 Brigade also had a role to deploy to West Germany should the Warsaw Pact threaten aggression.

I took command of the Mortar Platoon and the rest of the year was taken up with one training exercise after another. By Christmas, the battalion had trained frequently on Salisbury Plain; at Sennybridge, in Wales; Otterburn, in Northumbria; Stamford, in Lincolnshire; and in West Germany.

The exercise in Germany was remarkable not so much for what happened during the exercise, but for what happened afterwards. For once the RAF had its transport aircraft in place, ready to take us home. Within hours of the exercise ending, I found myself in charge of about fifty passengers for a flight to RAF Lyneham, in Wiltshire, ensconced in the body of an *Argosy* transport plane. The *Argosy* was distinguished by the fact that its nose wheel lifted the body of the plane only two or three feet clear of the ground. Apart from its transport role, it was designed to drop paratroopers, so its rudimentary seating was arranged in long lines on either side of the aircraft.

Almost as soon as we took off the soldiers were asleep, having had virtually none for the past several days. I woke up to look down on Lyneham, but the aircraft did not make an approach. It circled the airfield repeatedly. The RAF's SNCO air quartermaster, whose job it was to look after passengers and cargo in the body of the plane, lifted a floor panel almost at my feet and disappeared below the decking with a large spanner and a hammer in his hands. There was a fair amount of banging before the air quartermaster reappeared. 'Problem?' I asked. No response and he went directly to the flight deck. He was back within seconds. 'The captain wants to speak to you, sir'. I got out of my seat and went to the flight deck. The aircraft's captain said that the nose wheel's warning light was on and the wheel appeared not to lock when lowered. It was very possible that it would collapse on landing and the plane would skid along the runway on its belly. He asked if the troops should be briefed, or would it be better just to attempt a landing, with them still in the land of nod. He was concerned about passenger panic.

I was entirely confident that there would be no panic and said I would brief the troops, but wanted to know what they were to do in the event of the nose wheel collapsing. The captain's instructions were that we should all adopt the brace position and remain like that as he attempted to slow the plane after landing. He would give the order for everyone to unbuckle their seat belts, stand on their seats, which provided a more unobstructed surface than the floor that had our personal equipment on it, turn to face the rear of the plane and shuffle their way to exit at the rear doors, just as paratroopers do in many war films. We were to take nothing with us. The troops were woken and I passed on the captain's instructions.

It was tense as we came in to land. I looked out of my window to see that most of RAF Lyneham's personnel had left their work places to watch. The aircraft landed gently, but within seconds the smell of burning rubber was obvious and black smoke coiled through the decking at our feet. The nose wheel was down, but it was rubbing on its housing. We felt the plane slew a bit, then its path was corrected. All the time we were slowing, the acrid smoke was building up. At last the captain's voice told us to unbuckle, stand up and begin our exit. The air quartermaster fired small charges that blew out both rear doors and while the plane was still coming to a halt, we jumped the two or three feet onto the runway. It was covered in foam and our plane was being chased by RAF fire engines that narrowly avoided running some of us down.

As I had made my way to the exit door, I had an official MOD briefcase in my hand. It contained classified documents that I could not leave behind. As I passed him, the air quartermaster saw the briefcase, snatched it from me, threw it back into the body of the plane and shouted that nothing was to be carried. The soldier behind me, sensibly, was not going to be delayed in reaching safety, so I was bundled out of the aircraft.

There was some delay until the fire officer declared the *Argosy* safe. Eventually, our kit and weapons were unloaded and delivered to us in the passenger terminal. Everything arrived save the briefcase. The RAF Police were called and the aircraft, luggage vehicles and luggage belt were thoroughly searched, without result. They decided to contact the air quartermaster, who, by this time, had returned home to his quarter. I went with the police and was very relieved when the air quartermaster produced my briefcase from behind his front door. It had been placed with the crew's luggage and the air quartermaster had only noticed it when the crew's bus had delivered him home. Flap over, but one way or another, it had been an afternoon to remember.

With all the exercises behind us, we had become adept at having all our weapons and equipment prepared for short notice, strategic moves by air and once we had got to wherever we were going, for onward tactical deployments by troop-carrying helicopters. Not content with that, the Ministry of Defence decided to add another string to our bow.

The battalion was selected to carry out a trial of a tactic called 'mobile penetration'. It was reasonably assumed that any future war in Europe would be chaotic. Front lines would be fluid and gaps would appear, presenting opportunities for small groups of soldiers to penetrate to rear areas to attack high value targets there, e.g. major headquarters and logistic installations. These 'soft' targets would be less well defended than units in the forward areas and the loss of such assets would place an enemy at a considerable disadvantage.

This tactic required a considerable re-think. The penetration might be up to 20 miles deep, the vast majority of that in enemy territory and it could take three to four days to patrol to the target area, before the attack and subsequent ex-filtration were attempted. It was obvious that we would have to be very fit to cope with the distances involved and the amount of ammunition, food and water that would be needed. Soldiers would be required to carry up to 70 pounds on their backs. A further problem was that our normal VHF radios did not have the range required for the distances involved. HF radios were necessary and that meant all signallers would have to learn Morse.

Trials began locally. Platoons were divided into two for the penetration phase. Each half platoon commander was briefed separately and the orders that he received were deliberately sparse. He was given only his route and a lie up area to be reached after the first night of penetration, and a frequency and a time to have his radio on, to confirm his position and to receive details of the next leg. The idea was to deny the enemy any information from any members of the battalion they might capture. 'I know nothing' was pretty much true. This procedure was repeated until the required degree of penetration had been achieved and then things changed. Instead of lie up areas for individual half platoons, we would find that we had arrived in a company assembly area. Then it was time to mount an attack and skedaddle back from whence we had come, again by half platoons.

By the end of the year the battalion had made a good fist of the penetration tactic, so much so that we were rewarded with exercises in Kenya and in France to continue the trial over different terrains and climates.

Kenya was my first experience of Africa and I was to spend six weeks there. I was part of the advance party and flew into Nairobi airport, but was quickly moved to the town of Nanyuki that lies at 6,000 feet, exactly on the equator and in the shadow of Mount Kenya. It was January 1968 and the battalion was one of the first to exercise in Kenya. A battalion of the Kenyan Rifles was stationed in Nanyuki, but there was no accommodation for us so we established a tented camp on the outskirts of the town. Since that time, the British Army has continued to use Nanyuki because it has access to such valuable training facilities and now a permanent, major training base is established there.

The advance party's task was to be ready to receive the main body of the battalion that was to arrive in about ten days' time. My particular role was to recce areas of bush that we could use for live firing exercises and to help write tactical exercises for my company. Every morning I climbed aboard a Land Rover to accompany my company commander, Major Mike Duffy, and we would drive into the African plains to locate suitable areas for training. We were assisted by liaison officers and SNCOs from the Kenyan Rifles, whose local knowledge was invaluable and saved us a lot of time. During the recces we enjoyed what amounted to the most wonderful safaris. Wild animals and exotic birds were everywhere. Truly, it was like being in a David Attenborough documentary and we were being paid to be there.

The plains around Nanyuki were home to the Samburu tribe. They were nomadic people, continually on the move to find grazing for their cattle. When we decided on what real estate we needed and for how long, a political officer was informed and the tribes people were moved to safe areas and were paid for their trouble.

When the main body arrived, the company deployed to a farm, made available to us for training by its ex-pat owner. The 'farm' turned out to be a vast area of savannah and bush, barely farmed at all. The ex-pat was in residence. He was an American, from a very influential family. Silver framed pictures of him with members of the British royal family were very evident throughout his spacious, colonial bungalow. He made us very welcome and a week or so into our training, he invited the company's officers to dinner. We scrubbed up as best we could, but still looked fairly dusty from the day's exertions. On our arrival, he immediately instructed his Turkana servants to heat some bath water for us, so we would at least be clean to sit down at his table. We had our baths in strict order of seniority, Mike Duffy first, Paul McMillen next and I was last. Paul called out that he had finished in the bathroom and, as he passed me in a corridor, he whispered, 'I think the old sod was spying on me!' It transpired to be true. Our host was gay and, in the 1950s, his family had banished him to this remote part of Kenya for the sake of propriety – attitudes towards homosexuality were vastly different then from those we hold today. The bath water was fairly murky and not all that warm by the time I got to it, but it was welcome nevertheless. Whether or not he spied on me I am not sure – and care much less.

Dinner went on and on. I was very concerned because I had to lead my platoon some 30 miles through the bush the following day to a place called Crocodile Falls. But the hours drifted by, helped by copious amounts of good wine and engaging conversation. It was nearly dawn by the time we returned to our camp. I had two hours before we were to set off.

Despite my best efforts, it was clear to everyone that I had enjoyed a long night. At first, I marched with the body of the platoon, while a sergeant from the Guards Independent Parachute Company, a specialist in bush work, led the way. It got very hot once the sun came up and I sweated out the alcohol after only a couple of hours' strenuous exertion, so I was able to assume my rightful place in the lead. We had some scary encounters with buffalo and, particularly, with hippos when crossing a fairly deep river, but we made Crocodile Falls by nightfall, very pleased with our effort.

We 'shot for the pot' to supplement our rations and the fishing in the river was supposed to be good. However, we did not have time for the rod and line stuff – we used thunderflashes (powerful, hand-thrown pyrotechnics that explode just as well in water as they do on land). A line of soldiers was stretched across the river, safely downstream from where two NCOs tossed the thunderflashes into the water. The explosions concussed the fish and the downstream soldiers merely picked them up as they floated by. However, one of the fish catchers saw a chance for some fun. He yelled 'Grenade!' The exit from the water was instantaneous and chaotic. Everyone was thoroughly soaked. The miscreant was immediately identified and made even wetter than the rest of us. That night the cook made an excellent fish supper. Such antics might be frowned upon, but they can help develop team spirit and raise morale. They certainly set soldiering apart from other professions.

The companies rotated through several areas offering very different terrain. After spending some time learning bush craft and practising navigation and patrol techniques on the plains, we moved to the lower slopes of Mount Kenya. The mountain is 17,300 feet high. It has a tree and bamboo belt that extends from roughly 8,000 feet to 11,000 feet. In this belt we were able to gain some real jungle experience. Navigation below the dense forest canopy relied entirely on compass work and counting paces to know just how far you had travelled. Movement was often difficult, particularly in bamboo. The humidity was high and a myriad of insects and leeches seemed to have a taste for British blood. At night, the noisy calls of wild animals added to the strangeness of the environment, but no one could deny the excitement of just being there. However, some liked it more than others.

We had established a jungle camp. Four-man tents were erected to accommodate the soldiers, with bigger shelters in the cooking area. One officer had a phobia of snakes. The three of us, who shared a tent with him, had returned from our night exercises, but he was still in the jungle. Our sleeping bags were stretched out on canvas camp beds, but we only used the liners to sleep in. We cut a length of jungle vine, curled it up, put it at the bottom of his liner ... and waited. Eventually, he returned. It was pitch black. We pretended to be asleep. We heard him get undressed; sit on his camp bed and prepare to lie down. As his feet touched the vine, he screamed and did something quite remarkable. It can only be described as akin to toothpaste being squirted out of a tube, if its sealed end was hit with a hammer. In one movement, not only did he get out of his liner, but managed to clear the top of his camp bed. We all tried to replicate his acrobatics in the morning, but without anything to provide purchase, we could only manage to move a few inches.

Our cook also had a camp bed experience. He slept next to the cook tent. One night he awoke to hear someone rummaging around near his meat safe. 'Oi, what are you doing?' he shouted. A huge, feline roar answered him and, lucky for him, the leopard disappeared back into the jungle!

Life in Tidworth was fairly mundane compared to the excitement of Kenya. However, France was on the horizon and in August 1968 the battalion moved by ferry and troop train to Camp Valdahon, some 20 miles from Besançon, quite close to the border with Switzerland.

We were assigned barrack accommodation and our quartermaster was the first to discover that the French Army had a different view of life from our own. He had taken over our accommodation and was dealing with his French opposite number over things like rations, meal timings and cleaning materials. Towards the end of his negotiations, he asked what the designated day was for soldiers to exchange their used bed linen. The Frenchman looked up incredulously.

'You are only here for three weeks,' he said, 'Why do you want a sheet exchange?' *Vive la difference!*

The camp was home to a Chasseur regiment, equipped with France's latest AMX30 tanks. The regiment was extremely hospitable, but we were not to exercise with them. We continued the penetration trial in mountainous, largely forested country. The Valdahon training area was not large and for the final exercise, we were dropped off well away from it, in our half platoons, deep in the local countryside. The exercise plan was for us to patrol into the training area and mount an attack there. The local population and gendarmerie had been co-opted to be on the lookout for us, so it was a challenging task. During the three weeks that we were at Valdahon, it rained incessantly, but it saved the worst till last. The heavens opened for that exercise and we were drenched from start to finish. It helped us somewhat, as it kept most of the local population indoors.

Having a civilian 'enemy' was novel and heightened the degree of difficulty involved in undetected movement. But there was an element that nobody had foreseen. During World War II, this part of France was Maquis country. Many of the older generation were entirely game to help soldiers, who, in their view, were escaping from the gendarmerie. One of our patrols was hidden by an old lady, who flatly denied to the gendarmerie that she had seen any British soldiers, while all the time she had them concealed in her house, in hiding places last used by the Maquis. She said that she had not had such fun in years!

We always moved at night, but we were used to two major problems when concealing ourselves during daylight hours. However deep we hid ourselves in dense foliage that was always remote from human habitation, cows would frequently find us and gather round, eventually bringing farmers to investigate. The other problem was courting couples! No further comment required, other than that the rain did not put them off.

While we were in France, tension spilled over in Czechoslovakia. Troops from the Soviet Union, Poland, Bulgaria, the DDR and Hungary combined to invade the country and brought to a halt Alexander Dubček's liberal reforms that he had made earlier in the year. The invasion caused 137 people to be killed, with some 500 injured. It brought widespread condemnation, not only from western democracies, but from some communist countries too, most notably China. It would have repercussions for us in the months to come.

During the exercise in France, I had turned an ankle badly. Over the past two years, turned ankles had reoccurred several times and the injury was becoming more and more frequent. I now had my ankle in a cast and hobbled around on crutches. As a last resort I was sent to the RAF's rehabilitation hospital at Headley Court, in Surrey, to see what could be done about strengthening my ankles because, without improvement, my infantry career was threatened – if I was unable to run, jump and march, then I would be unable to serve.

RAF Headley Court was established in World War II to rehabilitate injured air crew. When as much as could be done medically to heal their wounds, downed air crew would be sent to Headley Court to begin a programme of intense physiotherapy. The staff there had the reputation of giving up on nothing and, by the time I went there in 1968, it had established a world-wide reputation as a centre of excellence. Its patients might be a bit battered and sore, but no quarter was given. Individual exercise programmes were designed and however much some of the exercises hurt, you were expected to persevere. There was one cardinal rule – no patient was allowed to touch, let alone help another, unless a member of staff asked them to do so. Failure to observe this rule brought severe consequences.

The hospital took officers and SNCOs from all three Services and also members of their immediate families. Along with every other male patient, I had noticed the very attractive daughter of an RAF officer. Very sadly, at a young age, she was suffering from Parkinson's disease. She had difficulty in walking and sometimes would run out of momentum completely. One mid-morning, classes had been suspended for coffee break. The girl was crossing the gymnasium in front of me when she came to a stop. However hard she tried, she could not get her legs to move again. Here was a chance to meet her. I walked casually by and nudged her shoulder. It was just enough to allow her to move. As she turned to thank me, a hand was placed on my shoulder. A SNCO physio had seen what had happened.

'You know what's coming, sir – 1,000 press ups. In your own time.'

The press ups took hours and were done publicly to make the point to others that the rule was to be properly observed. I could take as many breaks as I liked and had to, because eventually the build-up of lactic acid in my arms made completing even a few repetitions extremely painful.

The rehabilitation programme was mixed with a lot of fun. Every Wednesday afternoon an old RAF bus would take us to London's West End to see a matinee performance of a top show. The tickets were provided free by the theatres as an appreciation of the service of the injured patients. Also, once a week there was a disco, held in the very attractive, wood-panelled officers' mess. Nurses from local hospitals in Leatherhead and Epsom provided our partners. Dancing was a bit difficult for those of us on crutches. The chaps in wheelchairs were the lucky ones – the girls sat on their laps!

I spent two months at the hospital. It was hard work. I was eventually put in a 'black' class of one – me. I had to complete two, quite long, daily, cross country runs over the Epsom race course; had hydro-therapy classes; and a lot of gym work. My passing out test was to stand, bare footed, on a medicine ball for a couple of hours. Again I could take breaks, but I had to try and stand on the ball for twenty to thirty minutes at a time. The small muscles in my ankles had to work overtime for me to retain my balance. It might sound easy, but it was far from that. Without exaggeration, the expertise of the staff and the work done at Headley Court enabled me to continue to serve and to complete a full career. I owe them a lot.

The penetration trial continued to show promise and the MOD decided that it should be elevated to brigade level. To achieve this the battalion was put under command of 16 Parachute Brigade, based in Aldershot, and two of the Parachute Regiment's battalions would also join the trial. However, for the time being the Pompadours would remain in Tidworth.

We were still there at the turn of the year, when, in quick succession, we exercised in Denmark/Germany and in Iceland. We were issued with winter warfare clothing and needed it, especially in Denmark and Germany where the temperatures were extremely low and snow and ice made living in the field very challenging.

The exercise in Denmark was a NATO response to the Warsaw Pact's invasion of Czechoslovakia earlier in 1968. The battalion was flown into Denmark, to stage there and meet up with elements of the Danish Army, before flying south to the Kiel area of West Germany. There units of three nations would exercise together, as the Danes and Brits joined armoured units of the Bundeswehr. It was a demonstration of western capability and common resolve. The Soviets took note and heightened the degree of readiness of their forces in East

Germany. They also deployed some signal intelligence units to work against us during the exercise. The exercise was in the border area with the DDR and we and the Soviets were only a matter of a few miles apart.

As the mortar officer, I worked in battalion headquarters when we deployed in the field. Several days into the exercise I was manning the battalion's command net to the companies, when I recognised the voices of other colleagues and then my own voice in the radio traffic, but the conversations were those that had been held the previous day. It was confusing and difficult to discern which messages were real and which were bogus. The Soviets had intercepted our previous day's radio traffic, recorded it and were now rebroadcasting it intermittently on the current day's frequency. They were letting us know that they were there and were giving us a glimpse of their ability to disrupt. We had procedures to change frequencies in such an event and did so repeatedly over the next few days. It was a very tense time in Europe. Fortunately, radio intercept was the closest brush we had with the Soviets, but we were to remain eyeball to eyeball with them for a long while yet. Glasnost and a thawing of relationships with Warsaw Pact countries, was still two decades away.

Less than a month after leaving West Germany, the battalion had moved north, a long way north. We were in Iceland, the first British troops to be there since the end of World War II. Our reception was mixed. Some Icelanders welcomed us; most were indifferent; and a vocal minority opposed and contested our presence in their country, despite its membership of NATO.

Because training in Iceland was a first for the British Army, a detailed reconnaissance was required. The CO, still Lieutenant Colonel John Dymoke, selected me to be the exercise's adjutant and I was to accompany him on his recce. We flew to the US naval base at Keflavik, in the extreme west of the country and it was from there that we would mount our training exercises.

The British embassy, in Reykjavik, had made great play of the fact that John Dymoke was the Queen's Champion, an honorary position that his family had held for centuries. The coronation service is like a wedding ceremony in many ways, one of which is that at a particular stage in the service, the Archbishop of Canterbury asks if anyone has an objection to the future king, or queen, becoming the monarch. At that point the Queen's Champion steps forward and drops a gauntlet to the floor, enacting an age-old challenge. This quaint custom and Colonel John's appointment, attracted the Icelandic media's attention and their coverage brought a heightened awareness of the imminent arrival of British troops amongst the civilian population. It also fascinated the Americans.

On landing in Keflavik, the CO and I were shown to bachelor officers' quarters where we had adjoining rooms. Soon after settling in, there was a knock at my bedroom door and the CO explained that he had a small crisis. He thought he had packed a suit, but on taking out the suit's hanger from his case, he discovered that he had forgotten his trousers. He was a tall man and asked me to see if I could find an American officer, of about his size, willing to loan a pair of dark grey strides. I did my best, but the trousers I came up with were about three inches too short in the leg. The CO's choice was between the cavalry twill trousers that he had travelled in, or appearing in charcoal grey, with all the sartorial elegance of 'Coco the clown'. Sportingly, he chose the clown option and got some strange looks from our American hosts that evening. They also took an unusually close interest in our eating habits. They had specifically included peas on the menu, to see us use the back of the fork to eat them. It was going to be a steep learning curve for both sides.

During the recce we were surprised by three things:

- How isolated the American forces were, being very largely confined to their base. They could leave it, but an Icelandic police post was at its entrance and it was very like crossing an international border to reach Iceland's hinterland.
- The number of Russian-made vehicles that were in evidence and the strident antipathy towards NATO of an active minority of Icelanders.
- The sheer desolation of the interior over which we would exercise. We learned that the Americans had used some of the same areas to trial their moon landing vehicles.

Our recce was conducted at the tail end of winter when all was still ice and snow. Much of that had gone by the time the battalion arrived a few weeks later. We brought the Band and Drums with us so that they could perform at a number of venues and generate good relationships with the civilian audiences. Meanwhile, the rest of the battalion deployed to the centre of the country, remote from settlements of any size. Our training would have a minimal effect on the environment, but the environment would have a significant impact on us. The surface is pumice rock, which is sharp and extremely abrasive. We set about conducting a series of patrol exercises over deeply fissured, open country, often across the terminal moraine of glaciers. Cuts and abrasions were commonplace, but it was our boots that really suffered. They were cut to shreds in a matter of days. The Quartermaster had brought a reasonable number of replacements with him, but nearly everyone needed boots replacing and more had to be flown in for the exercise to be completed.

The local press had not forgotten the Queen's Champion story and ran it again when the battalion arrived. Sometime after that, when training was in full swing, battalion headquarters had established a tented camp at the foot of a glacier, miles from anywhere. We were quietly going about our business when we saw three buses making their way slowly up the dirt track that terminated about 600m to 700m from the camp. People got out of the buses, formed up, unrolled banners and made their way towards us. The banners made clear that they wanted us off their island as soon as possible. I informed the CO and was told to go and meet them. I did so at the edge of the camp. An elderly gentleman stepped forward and introduced himself as a medical doctor and the de facto leader of the group. He asked to see the CO. I told him that was not possible. He conferred with several of his people and then handed me a letter addressed to Lieutenant Colonel Dymoke. I took it; asked him to wait; and went back to the CO's tent. The CO opened the letter and read it.

'How big's this bloke?' he asked.

'About my size,' I replied.

'Is he old or young?'

'I would say the wrong side of sixty,' I guessed.

'Good,' said the CO, 'he's challenged me to a duel!'

We had a hurried discussion and then the CO said, 'Tell him that this is a challenge to the Queen and it will have to be sent to London.'

I duly conveyed the message. The doctor's response was that he would accept no delaying tactic and expected us to take action. He added that he had sent a copy of the letter to the Icelandic media. In doing so he had made sure that we could not ignore it, or hush it up.

After walking around the perimeter of the camp a couple of times, shouting at us to go home, the group of protesters lost interest, made their way back to the buses and left. I was sent to the embassy in Reykjavik to seek guidance.

The diplomats there were non-plussed and the contents of the letter were hurriedly sent to London. By the following morning we had a response. The staff at Buckingham Palace had come up trumps. Armed with their advice, an announcement was made by a diplomat from the embassy. The diplomat regretted the opposition to British troops exercising in Iceland. He emphasized the country's membership of NATO and the security benefits that came with it. Turning to to the doctor's challenge, he said that the British regarded it as invalid. The doctor was a commoner and, as such, he had not the standing to challenge a royal personage. The matter was closed. Phew!

Earlier in this chapter I mentioned that the battalion had been placed under command of 16 Parachute Brigade, but that it would remain in Tidworth for the time being. However, shortly after returning from Iceland, we were ordered to co-locate with the rest of the brigade in Aldershot and another unit move had to be undertaken. The move would come just two years after arriving in Tidworth from Berlin.

Previously, I have pointed out how difficult moves were for wives and families. This next move would be no different. Many of the quarters in Aldershot were new, but they were poorly constructed and notwithstanding that they were only a few years old, the need for significant repair was beginning to emerge. The problems of children changing schools are present in any move, but now the battalion had been in England for two years, a further problem arose. Bright, go-ahead people tend to marry similar partners. Many of the wives, particularly those with more senior children, who did not require constant child care, had found themselves jobs, even in Tidworth's comparative remoteness. Those jobs had to be sacrificed and the incomes they attracted had to be surrendered, if the families were to stay together. All that came on top of two years when the battalion had trained abroad on five separate occasions, in addition to the training conducted in the UK. That heavy training schedule would have required most of the married men to be away from home for about nine of the twenty-four months that the battalion had been in Tidworth. And that was not the full extent of the separation the families suffered. Once on the promotion ladder, most NCOs attend career courses that are held centrally, which can last from a couple of weeks to three or four months, resulting in yet more separation. Generally, the married men thoroughly enjoyed the challenge and variety that training brings, but it is a

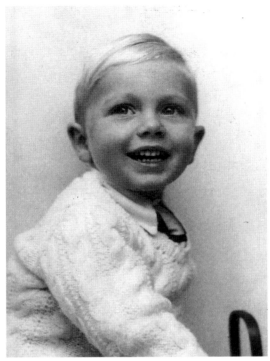

My first portrait, aged three.
(Donald J. Donovan, Frinton-on-Sea.)

Lower VI Form, Atherstone Grammar School, June 1961. Sue Waterhouse – front row, extreme right. Colin Groves – second row, 3rd from the right.

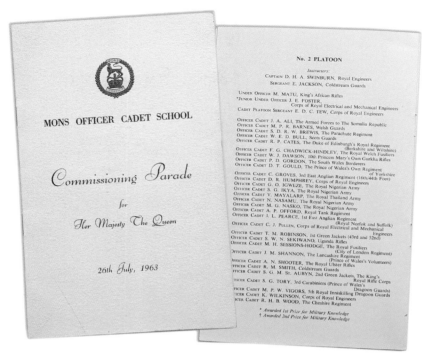

MONS OFFICER CADET SCHOOL

Commissioning Parade

for

Her Majesty The Queen

26th July, 1963

No. 2 PLATOON

Instructors:
CAPTAIN D. H. A. SWINBURN, Royal Engineers
SERGEANT E. JACKSON, Coldstream Guards

UNDER OFFICER M. MATU, King's African Rifles
*JUNIOR UNDER OFFICER J. E. FOSTER,
    Corps of Royal Electrical and Mechanical Engineers
CADET PLATOON SERGEANT E. D. C. TEW, Corps of Royal Engineers

OFFICER CADET J. A. ALI, The Armed Forces to The Somalia Republic
OFFICER CADET M. P. R. BARNES, Welsh Guards
OFFICER CADET S. D. R. W. BREWIS, The Parachute Regiment
OFFICER CADET W. E. D. BULL, Scots Guards
OFFICER CADET R. P. CATES, The Duke of Edinburgh's Royal Regiment
    (Berkshire and Wiltshire)
OFFICER CADET F. G. CHADWICK-HINDLEY, The Royal Welch Fusiliers
OFFICER CADET W. J. DAWSON, 10th Princess Mary's Own Gurkha Rifles
OFFICER CADET P. D. GORDON, The South Wales Borderers
OFFICER CADET D. T. GOULD, The Prince of Wales's Own Regiment
    of Yorkshire
OFFICER CADET C. GROVES, 3rd East Anglian Regiment (16th/44th Foot)
OFFICER CADET D. R. HUMPHREY, Corps of Royal Engineers
OFFICER CADET G. O. IGWEZE, The Royal Nigerian Army
OFFICER CADET S. G. IKYA, The Royal Nigerian Army
OFFICER CADET V. MAYALARP, The Royal Thailand Army
OFFICER CADET N. NASAMU, The Royal Nigerian Army
OFFICER CADET M. G. NASKO, The Royal Nigerian Army
OFFICER CADET A. P. OFFORD, Royal Tank Regiment
OFFICER CADET I. L. PEARCE, 1st East Anglian Regiment
    (Royal Norfolk and Suffolk)
OFFICER CADET C. J. PULLEN, Corps of Royal Electrical and Mechanical
    Engineers
OFFICER CADET T. M. ROBINSON, 1st Green Jackets (43rd and 52nd)
OFFICER CADET S. W. N. SEKIWANO, Uganda Rifles
OFFICER CADET M. H. SESSIONS-HODGE, The Royal Fusiliers
    (City of London Regiment)
OFFICER CADET J. M. SHANNON, The Lancashire Regiment
    (Prince of Wales's Volunteers)
OFFICER CADET A. N. SHOOTER, The Royal Ulster Rifles
OFFICER CADET R. M. SMITH, Coldstream Guards
OFFICER CADET S. G. M. St. AUBYN, 2nd Green Jackets, The King's
    Royal Rifle Corps
OFFICER CADET S. G. TORY, 3rd Carabiniers (Prince of Wales's
    Dragoon Guards)
OFFICER CADET M. P. W. VIGORS, 5th Royal Inniskilling Dragoon Guards
OFFICER CADET K. WILKINSON, Corps of Royal Engineers
OFFICER CADET R. H. B. WOOD, The Cheshire Regiment

* Awarded 1st Prize for Military Knowledge
† Awarded 2nd Prize for Military Knowledge

Commissioning Parade Programme listing No 2 Platoon members showing their regiments and overseas armies.

Lieutenant Jeremy Veitch indicates the tented accommodation at Radfan Camp, Aden in 1966.

Commanding a patrol in Sheikh Othman, Aden
(second man back). The steel helmets indicate a
period of rioting was in progress.

About to run the anchor leg of the 4 x 200m relay in the Aden Inter-Unit Athletic
Championships, 1967. Taking over from Private Jardine.

Sue. One of the photographs that fell from a torn envelope at an Aden checkpoint. May 1967.

Rudimentary mine clearance hook designed to snag any command wire connected to an explosive device. Aden 1966–67.

Aden checkpoint.

Relaxing with members of the Mortar Platoon in a bivouac area in Kenya, January 1968

'Under the Blades' – Sue and I leave the church on our wedding day,
10 September 1969. Lieutenant Pat Shervington front left.
Lieutenant Paul McMillen front right.

'Bleak House'. Our first quarter in Warminster, 1969. A generous size and fully
equipped, but it was a totally uninsulated, concrete fridge in winter.

The Adjutant. In my office in Paderborn with the Pompadours fly out state behind me. March 1972.

Meeting Rt Hon Willie Whitelaw, Secretary of for Northern Ireland outside battalion headquarters in Hastings Street (a commandeered mill) with Major John Woodisse. Belfast tour 1972.

No Premier League club has so much firepower! The Pompadours football team, with their trophies, at Fort George, Londonderry 1973. Foreground kneeling – left Sergeant Robbie Allen (team captain), right Captain Jimmy Jenks MM (manager).

With Captain Mark Phillips on his first day at Sandhurst. My platoon sergeant, Colour Sergeant Derek Rossie, Grenadier Guards third from right. March 1974.

(Colin Bowerman, Cassidy and Leigh, Guildford)

different story for their wives. It is not surprising that from about this time, in the late 1960s, the divorce rate in the Army began to increase and, so too, did the number of families who elected to buy their own homes and for the wives and children to settle there, while the husband served away from them. It placed stability above family togetherness and in doing so, it also placed its own strains on marriage.

I moved with the battalion to Aldershot, but regrettably had to give up command of the Mortar Platoon. The battalion was warned of another overseas commitment, this time a tour of six months as part of the United Nations' peacekeeping force in Cyprus, beginning in November. I was not to go with them, so the Mortars needed a commander who would accompany them. I had been selected to instruct on the Skill at Arms Division at the School of Infantry in Warminster – so it would be back to Salisbury Plain in September. To fill the few months I would spend in Aldershot profitably, I was made Intelligence Officer (IO) and Assistant Adjutant – basically, desk jobs. I did not mind the IO bit, as I had to research the situation in Cyprus and produce briefing materials to supplement those we received from the MOD. It was interesting and it contrasted with the assistant adjutant's role.

The adjutant is, arguably, the hardest worked officer in the battalion. Apart from running the CO's office, he is also the principal human resources officer, dealing with everything from promotions to courts martial, and provides the point of first contact for a great deal of the mail that comes into a battalion daily. Whoever occupied the appointment always needed help and the assistant adjutant provided it, picking up anything the adjutant could not find time to address.

It was one of those 'no time to look at that' tasks that came my way and kept me fully occupied. The Queen Mother was the regiment's Colonel-in-Chief, but, in addition, each battalion had its own Deputy Colonel-in-Chief. Ours was Princess Margaret. She had been due to visit the battalion in October of the previous year, but a very last minute indisposition had prevented the visit from taking place. That disappointment was to be rectified and another visit was arranged for her in June 1969. The Army Show was to take place shortly afterwards, again in Aldershot, and someone had the bright idea of aligning both commitments to reduce the work involved.

The battalion was required to work up two very different arena acts for the

Army show, on a scale bigger than those seen at the Royal Tournament. We could also utilise them for the royal visit and thus, avoid the hours of practice needed to polish a ceremonial parade.

The first was a helicopter raid on an 'enemy', using assault and troop carrying helicopters. The spectacle was made all the more impressive with the expenditure of a considerable amount of blank ammunition, smoke and thunderflashes. The second act was a re-enactment of the battle of Salamanca, part of the Peninsular Wars, fought in 1812. At this battle a forebearer regiment captured a French eagle, which, over 150 years later, was the reason for the embarrassment of the French director at Spandau Prison. French eagles are the equivalent of British colours. Few were ever taken, but the Salamanca eagle is still in our possession and remains extremely important to us. I was to be the arena commentator for both acts and had to write suitable scripts.

Both events drew large crowds and the acts were dramatic in their different ways. The helicopter assault was all action man stuff, while the early nineteenth-century battle had great colour and the dash of a 'French' cavalry charge, repulsed by the fortitude of the gallant Redcoats. My short time as a commentator was fun, but I did not expect, or deserve, to be headhunted by the national broadcasting companies.

In the middle of preparing to deploy to Cyprus, the battalion had to undertake street lining duties, in London, for the state visit of the President of Finland. That meant masses of drill in order to stand in a gutter for ages, waiting to salute important people as they made their elegant way past. It was not our idea of fun. Perhaps, for once, we had some sympathy with the Guards' regiments and the ceremonial role they perform as a matter of course. Interestingly, the Queen and the President were cheered as they went by. A rolling swell of loud booing warned us that Prime Minister Harold Wilson was on his way.

My swan song, before my posting to Warminster, was to carry the Queen's Colour at a parade for the Regiment to receive the Freedom of Hertford, in the presence of the Queen Mother. It is every subaltern's right to carry a colour, if the opportunity presents itself. The Colours are the heart of a unit. They are works of art in their own right, created by the Royal School of Needlework, with beautiful, intricate embroidery overlaying the silk, which provides the body of the Colours. They are consecrated on being taken into service and never leave their unit's possession. Outsiders can look, but may not touch. It was a great honour to carry one on parade and now it was my turn, if I could convince the adjutant to give me my chance. The adjutant, Kerry Woodrow, was sympathetic, provided I could find six men, of similar stature, to form a colour party. It was not too difficult.

I needed to find a subaltern junior to me, if I was to carry the Queen's Colour, which was the job that goes to the senior chap. There was an answer to both the stature and seniority problems. I am 5 feet 8½ inches tall. Peter Dixon, who was commissioned after me and became a particularly close friend, is 5 feet and 8 inches short, although for decades he has erroneously claimed my statistics should be reversed. He was keen to carry the Regimental Colour. With him on board, I had only to find the four members of the escort. I consulted the RSM and quickly had a warrant officer, two colour sergeants and a drummer lined up – job done.

The main body of troops on the parade and the parade commander came from our 5th Battalion (Territorial Army soldiers). We provided the Colour Party and the Band and Drums. The venue was a hockey pitch of the Richard Hale Grammar School in Hertford. It was an attractive setting, but the arena was somewhat restricted. When the parade had to present arms for a royal salute to greet the Queen Mother, she was on a saluting dais only about 30 feet from me.

At that stage, before the Freedom had been conferred, only the Queen's Colour was on parade. A stiff breeze was blowing, but occasional stronger gusts threatened to make life difficult for me. The parade commander ordered 'Royal Salute'. That was the moment for me to let the Colour fly. Up to then I had it firmly under control, with a bottom corner gripped between my right hand and the pike (pole) on which it was mounted. As I let the Colour fly, a strong gust came out of nowhere. The Colour made a sharp 'crack' as the wind whipped it to its full extent. I had to struggle to remain upright and stationary. The next words of command were 'Present Arms'. A royal salute, with a Colour, requires the ensign to extend his right arm fully to the two o'clock position, releasing the base of the pike from a socket in the Colour belt. At that point the Colour and its pike are held only by the right hand and it remains that way until the Colour is lowered to the ground in a sweeping arc and the arm is drawn back into the body. In the gusting wind, it was extremely difficult for me to make the salute the elegant movement it should be, so much so that one of the

colour sergeants said, *sotto voce*, 'Hang on to the fucking thing, sir.' I hung on
and the rest of the proceedings passed off without further incident.

When all the formal ceremonies were over, I was presented to the Queen
Mother and we exchanged a few pleasantries.

Then she said, 'It was a very trying day for carrying Colours, Mr Groves.'

'Yes Ma'am,' I replied, 'the breeze was tricky.'

She smiled a knowing smile, 'I did enjoy your colour sergeant encouraging
you!'

She moved on.

Just a few weeks before I was due to leave the battalion, the COs changed.
John Dymoke left and in came an officer with a 3rd Battalion background
and a reputation, Lieutenant Colonel Keith Burch. The 'new broom'
immediately swept with a ruthless vigour. He was remarkable in his grasp
of detail, military knowledge and his diligence, but appeared devoid of the
touch with people that is essential for a commanding officer to succeed. I
was glad to be leaving.

All the while I was in Tidworth and Aldershot, I saw as much as I could
of Sue and many weekends were spent in London. However, the frequent
training deployments meant that I was often away for several weeks on end
and this did little to promote the Army in Sue's estimation. I took whatever
opportunities came along to introduce her to my friends and to the life I
led. She really was starting from scratch with the Army, so much so that she
bought an illustrated Ladybird book that explained the rank system and its
badges. She attended her first military ball in Tidworth, stunning in a new
pink gown, long gloves and expensive hair-do, but suffering some inevitable
nerves. She went to another ball in Aldershot and one or two parades,
however, she never became a fan of marching men and military music. The
event that made the most lasting impression on her was an extraordinary
boating excursion over one Whitsun weekend, when one of our number,
Trevor Veitch, persuaded three more of us to bring our girlfriends for a
two-night excursion on the Thames. A luxury river cruiser was promised
at a knock-down price. *Chakdina* turned out to be a very small boat, with a
half roof, just capable of squeezing eight people on board. It was normally
hired by the hour! The surprising thing was that nobody 'jumped ship' and
that thirty-six hours after setting off, we were still friends and had had some

remarkable experiences along the way, possibly enough to fill another book.

By now Sue and I knew that we wanted to get married, but it was not as simple as that. In 1969 any soldier was, of course, free to marry, but, and it was a big 'but', the Army only recognised marriages when soldiers reached twenty-one years of age: for officers the age limit was twenty-five. Recognition brought with it marriage allowance and the right to be allocated an Army quarter. Officers and soldiers under the age limits had to make their own way until those limits were reached. Without Army support, married life was financially impossible for everyone, save the very few who had private incomes. However, in February 1969, I had reached my 25th birthday and Sue made it clear that she was getting fed up of waiting. I got an ultimatum. Unless we were married before Sue was twenty-six, she was off. We set a date – 9 September 1969 – by then her birthday was all of twenty-seven days away – these things should not be rushed!

Army formalities were not quite over. Before our engagement could be announced, I was required to write to the Colonel of the Regiment asking his permission to marry Sue. The letter had to be very formal. It began:

'Sir,
I have the honour to request your permission to marry Miss Susan Waterhouse of Dordon, Warwickshire ...'

At the time, the Colonel of the Regiment was Lieutenant General Sir Richard Goodwin. He was a pleasant and sensible man and wrote back along the lines that he had probably only met me twice and had no knowledge at all of Miss Susan Waterhouse. Of course we could get married and he wished us every happiness. He also wrote that this outdated custom had no place in a forward-looking regiment and should be discontinued immediately. I was the last Royal Anglian officer required to jump that particular hurdle and the announcement of our engagement was made in the national press, which was the custom for officers at the time.

My first tour of regimental duty commanding soldiers, drew to an end after some six years of helter skelter experience. I had grown up. I was ready for marriage to Sue and for the next phase of my life. I was to be promoted to acting captain, to be an instructor at a centre of excellence and for the next two years I would work with SNCOs of an elite Corps. The personal and professional changes would be enormous.

CHAPTER SEVEN
STARTING LIFE TOGETHER
IN WARMINSTER

DURING THE MONTHS before I left the battalion, there were wedding plans to make. Sue did the lion's share, but I contributed where I could. I wanted – and we agreed on – a military wedding, which have their own colour and not a little glamour. None of our parents were particularly affluent and as we did not want to put a strain on their finances, we decided on a small event, to be held in the mid-week.

We were similarly financially stretched, so we had to make some compromises. The first of those was the engagement ring. London spoilt us for choice, but any ring that we liked was hopelessly out of our price range, so Sue happily opted for a second hand ring. We found a particularly pretty one and a platinum wedding ring to complement it.

The venue was not difficult to decide. As a teenager, Sue had been a Sunday school teacher at All Saints Church in Grendon, in Warwickshire, the next village to where her parents lived, in Dordon. The church, originally built in the 12th century, was located a little way outside the village, in archetypal English countryside. It provided an ideal setting.

My best man was Jerry Steele, a fellow subaltern, sportsman and great mate. We had served together since the battalion returned from Aden and had become staunch friends. On the day of the wedding we met up at the reception venue, the Chase Hotel, near Nuneaton, some seven miles from the church. We had a drink and got changed into ceremonial uniform – No 1 dress tunic (Blues), mess dress trousers (also blue), medals, woven gold epaulettes and sword tassels, crimson sash belts, chromed sword scabbards and white gloves. It was the bride's day, but no officer likes to be totally eclipsed!

We set off for the church in an enormous, black, very old Rolls Royce and had gone a mile or so when Jerry began to fumble in his uniform's pockets. I told him to stop mucking about, but he really had left the ring in our room at the hotel. Timings would be a little bit out from now on.

The 'Roller' dropped us at the church and left to collect Sue, her father and her only bridesmaid, Miranda, her cousin. Before we took our places in the church, Jerry and I had time on our hands and it was good to talk to some of our families and guests, just over sixty in all, as they arrived in their finery – ladies looking wonderful and men in uniform, or in morning dress.

We had asked Eric Venn, our former music master at the Queen Elizabeth Grammar School, to play for us and his playing was magnificent. At last, he broke into Mendelssohn's wedding march. I did not turn to see Sue process. I wanted to be surprised as she appeared at my elbow. I was not disappointed. Her dress, made by her mother, was (now for the technical bit) short sleeved, empire line, in white kafta, with a bodice of white flowers that matched her head-dress. She was radiant and everything I would ever want. Words failed me and hoped my expression conveyed to her all that I felt brimming inside me.

The vicar began the service and soon was asking us to kneel. Aahh – I should explain that mess dress trousers are incredibly tight and are buckled under the soles of knee-length mess wellingtons that are worn inside them. The buckles ensure that there are no wrinkles in the trousers, meaning that there is no give whatsoever. I tucked my sword out of the way and lowered myself gingerly onto the kneeler, hoping against hope that there would be no tearing of seams. That little problem over, the rest of the service passed quickly. As I made my vows I knew that I meant them and I was equally certain of Sue's, as she made hers to me.

All too quickly I heard a shuffling at the back of the church as my fellow officers made their way outside. We signed the register and Nimrod, from Elgar's Enigma Variations, soared through the building. Married, we made our way through a happy, smiling, well-wishing congregation. As we reached the church door I heard our guard of honour come to the special sword salute that is reserved for weddings. I had been part of such guards many times for fellow officers, now it was the turn of my wife and I to walk under the blades.

Weddings in the 1960s and '70s were simple affairs when compared to the extravagance of many held today. Then the service was almost always held in the early afternoon, timed to allow a reception to follow. By late afternoon/early evening the reception was over and the bride and groom would leave, amongst lots of well-wishing and tomfoolery, to begin their honeymoon. The process was short and sweet, but that did not limit the enjoyment that the occasions generated. Our wedding followed this pattern. I remember Jerry making a clever speech and I responded. I cannot recall any of what I said, except that I promised, in front of all I held most dear, that I would cherish Sue, a word not included in the wedding vows. Despite the inevitable ebb and flow of our marriage of over more than half a century, it has not been a difficult promise to keep. Marrying Sue was the cleverest, luckiest, most sensible thing I have ever done.

Amongst a lot of cheering, tin cans tied to bumpers and a kipper on the manifold, Sue and I climbed into my cream-coloured MG Sprite, waved our goodbyes, and sped off to Matlock, in Derbyshire to begin our honeymoon. We had only ten days before I had to report for duty at the School of Infantry. I had planned for us to travel up the eastern side of England, tour Scotland and return, via the northwest coast, to Sue's parents' house, before travelling to our new home in Warminster.

The honeymoon went vaguely to plan. It was very, very enjoyable, but there were one or two surprises.

Our hotel in the Scottish highlands was comfortable, traditional, turreted and dour. The next youngest couple to us were at least two, if not three, decades our seniors. There was a lot of nudging and winking whenever we appeared. Perhaps they were re-living lost youth!

We had our first disagreement in Oban, on the Scottish west coast, where we were to spend just one night. Sue decided that a hotel would be a waste of money and B&B accommodation would suffice. I thought B&Bs were not appropriate for a honeymoon. She prevailed, but when we knocked at the door of the place we were to stay, it was opened by a clone of Rab C. Nesbitt – the alcoholic, unemployed, Glaswegian layabout and central character of a 1990's comedy series. The man in front of us had lank, unkempt, straggly hair; a roll-up cigarette in his mouth; his top was a less than pristine string vest; and his trousers were past their sell by date. Our first impression was 'We're not staying here', but he said something utterly unintelligible, grabbed our bags and disappeared inside the building. We had no option but to follow. Thankfully, the owner was the low spot. Our room was clean and comfortable enough and he served a great breakfast the following morning.

South of Oban, just off the Kintyre peninsular, is the Isle of Gigha. My Uncle Ken and Aunt Jessie had a croft on the island. I had been a page boy at their wedding in 1948, but had seen them only a couple of times since. It seemed somehow rude to pass within a few miles of them and not to call in. That apart, Gigha is a beautiful island and I knew Sue would enjoy seeing it, so we made the twenty-minute ferry crossing from Tayinloan. Jessie was charming and wonderfully welcoming. Her family came from the island. Ken, on the other hand, more than lived up to his reputation of being a really miserable so-in-so. Within twenty-four hours of our arriving, he made it clear to me that if Sue and I were to enjoy his hospitality, I would be expected to do some work on the croft. I did two days of hedging and ditching and that was enough. We left him to his curmudgeonly ways. Our thoughts were very much with Jessie, poor woman.

Our next stop was the Lake District. We arrived in pouring rain and after one night, with the forecast predicting days of the same appallingly wet weather, we decided to cut our losses and head for Sue's parents' home. We had plenty to do before leaving for Warminster.

The MG was groaning with mainly Sue's possessions, when we set out for Wiltshire. Her parents would have to bring the wedding presents at some later date. I had consigned my stuff to Warminster before I had left the battalion, in what was referred to as 'MFO' – short for 'Military Forwarding Organisation'. In those days, MFO was the only way personnel could get their possessions from one posting to the next. Servicemen and women were able to obtain wooden boxes from the quartermaster, assemble them and pack their paraphernalia in them. The dimensions of the box were approximately 90 × 60 × 70 cm. That was fine for single soldiers, but it placed severe limitations on married couples. Admittedly, the Army provided all the furniture, fittings and utensils required in a house, but everything was a standard issue. It was extremely frustrating for wives, who wanted to put their own stamp on their homes, to have the boxes limit them to coffee tables, lamps, ornaments, pictures and soft furnishings. Owning larger items of furniture was simply impractical, unless you were prepared to pay for it to be removed privately for inter-UK postings, or have it put into store when you served abroad.

The upside for us was that I had been allocated 82, Elm Hill, Warminster as our quarter, a detached, three-bedroomed house, with a garage and gardens both front and rear. It came equipped with every conceivable household item, from a double bed to egg cups. I paid rent, but it was markedly below a commercial rate. For a few months, until the salary structure changed, we even had a batwoman, who came to clean two or three times a week. Most newly married couples would have given their eye teeth for such luxury.

On arriving at the School of Infantry, we went directly to our quarter to meet the estate warden who would 'march us in'. Estate wardens were invariably retired warrant officers, or SNCOs, who continued to serve in a civilian capacity, to help maintain the Army's housing estates. They were responsible for 'march ins' and 'march outs' – dreaded terms for any service couple.

For estate wardens to accept them, quarters had to be left immaculate, to the standards of cleanliness that had been imposed on me at Mons. Every item on a quarter's inventory had to be laid out for counting and inspection. Couples went up in an estate warden's estimation if the items were laid out in the order they appeared on the inventory, but you were very clever if you could work out that 'brushes bass – quantity 1' (a long-handled yard brush), should be positioned next to 'brushes SHU – quantity 2' (short-handled utility brush, or scrubbing brush in standard English). And we never were able to work out why quarters, left immaculate on 'march out', were not quite so immaculate on 'march in'. Nevertheless, after an hour or so, the house was ours.

After reporting for duty, I had an interview with Major General Gilbert, the Director of Infantry and commandant at Warminster. He was keen to learn a little of my military experience, before he asked more about me. He wanted to know where I was to live. I told him our quarter was in Elm Hill. He looked me straight in the eye and said, 'You'll freeze your nuts off!' He was not wrong.

Our spacious quarter was made largely of pre-cast concrete slabs. The walls and roof were uninsulated and the single-glazed windows had iron frames. A small, coke-fired stove in the kitchen contrived to cover everything in smuts. It heated two tiny radiators, one in the dining room and the other in the hall. The large sitting room had a coal fire that could be damped down to burn all night, but it failed to produce any meaningful heat. Upstairs every bedroom and the bathroom had a two-bar electric fire screwed to a wall, about seven feet from the floor. That was the sum of the heating in the house. Elm Hill quarters were amazingly efficient, residential refrigerators!

We had moved into No 82 in September. By mid-October we noticed that most of our neighbours were moving out of the master into one of the smaller bedrooms. Additionally, they were attaching plastic sheeting across windows. It made for grossly inadequate insulation, but they were preparing for winter as best they could. We thought we would be wise to follow suit and were glad that we did.

Leaving quarters aside, I should explain that once officers reach the rank of captain, the profile of their careers changes. The early part of officers' careers is spent commanding soldiers, but with about six years' service behind them, they are sufficiently experienced to serve in staff appointments in headquarters, or to have a role in a major training establishment. Once that point is reached, from then on, staff or training appointments tend to alternate with tours at regimental duty. Until fairly recently, officers tours of duty were for two years – not long to learn a job and be effective in it, although the variation the system brought was a big plus. However, the short tours meant considerable disruption to family life.

The Skill at Arms Division was commanded by a major, assisted by three syndicate officers, all captains. I was to be one of those. Each syndicate comprised four classes for students, taught by members of the Small Arms School Corps (SASC). This elite corps of weapons instructors numbered only about 150 and had the minimum rank of sergeant. Recruits into the SASC were already NCOs from other branches of the Army, who had proven instructional ability. Even then they had to pass a strict probationary course. Consequently, the level of instruction was extraordinarily high and the syndicate officers had to reach that same level.

The 'bread and butter' course was one of eight weeks, for junior NCOs, who were destined to become weapon instructors in their own units. Other, shorter courses were run for members of the Territorial Army and later in my tour of duty, a sniper course was introduced. I wrote the sniper course and in the process of developing it, I became a sniper, which was highly unusual for an officer. It was a skill to be put to use in a later tour.

Syndicate officers were required to give lectures and provide the commentary for the many outside demonstrations that were part of every course. I was placed in the care of the Demonstrations Sergeant Major (DemSM). He was to improve my instructional ability before I was let loose on a course. I gave a lecture on 'The Theory and Penetration of Small Arms Fire' – not a subject for the uninitiated. It went well and the DemSM seemed reasonably pleased.

'I'd give you a high "B" grade for that, sir,' he said, then added, 'but you'll be doing it again until it's a high "A".'

He proceeded to reel off a long list of constructive criticisms that would demand my close attention and a lot of effort. I gave the lecture five more times before he was satisfied, notwithstanding that I had reached an 'A' grade by the third attempt. It was time well spent. Ever since then, public speaking has never held any fears for me and I was to be heavily involved in instruction throughout my career.

Sue had a new job, too. For some time before leaving London to marry me, she had moved from her first job in a library in Little Portland Street, to work in the Inner London Education Library, housed in the Greater London Council's vast offices, on the other side of the Thames from the Houses of Parliament. It was a huge library. Her career was advancing quickly and at the time of our marriage, she earned more than I did. Things changed when she got to Warminster, but she set a trend that she maintained throughout my service career. Sue was never content to stay at home. She would work whenever she could and that proved to be more often than not. She quickly found a job with Somerset County Council's libraries and became a librarian at Frome, albeit not at the level that she had reached in London.

My working life in Warminster was very predictable. Courses were programmed many months in advance and there were virtually no surprises of any consequence. Nevertheless, the work was extremely satisfying. JNCOs would arrive in Warminster at the beginning of their courses and for most, it would be the first time that they had been outside the family of their units. They were invariably nervous, daunted by the standards that they were expected to reach in only two months. The syndicate officers observed many of the lessons they taught and progress was reviewed week by week with their class instructors. The change in the students was heart-warming. For almost all, their knowledge and instructional expertise expanded exponentially, but it was their self confidence that showed the greatest improvement. We did not just produce extremely able weapons instructors, we provided them with an ingrained self-belief on which to build much wider, successful careers.

For two weeks in the summer the Division got a break from instruction. It moved, as an entity, to provide much of the organisational infrastructure for the Army's and the National Rifle Association's (NRA) shooting championships at the world famous ranges at Bisley, in Surrey. The NRA owns the complex at Bisley and its ranges are remarkable in their size and variation – the complex has hosted the Olympic Games' shooting competitions. The syndicate officers became the range officers for the biggest ranges where the most important competitions were held. The conditions for each competition are very tightly written and anything other than 'letter of the law' observance would result in objections. It was the range officers' job to ensure that things ran smoothly, but it had the reward of being able to observe marksmanship at its best, be it long range riflemen scoring bullseyes at 1,000m, or fifty machine gun pairs (each gun has a No.1 and a No.2), firing simultaneously on Century range – it combined an awe-inspiring sight with a unique, unmistakeable cacophony of chattering gunfire.

The summer of 1970 also brought a first long leave for Sue and me. I had not realised until then that I had married a frustrated travel agent. Sue had an urge to see the world that has not diminished over time. In 1970, the Sunday broadsheets carried advertisements depicting air conditioned, long wheel based Land Rovers traversing the desert, with *'Safari'* emblazoned on their sides. The compelling pictures promoted adventure holidays in Morocco and Sue wanted to go. I thought that I did a lot of what was on offer as my job, but I had never been to North Africa, so I was easily persuaded.

We flew from Gatwick by Dan Air, a cut price company with a fleet of elderly planes and ours rattled its way to Tangier. Coming in to land I looked down at the airport's terminal. On the approach road I could see four bright yellow lorries and trailers. As we descended, it was possible to make out *'Safari'*, written in red, on their sides. What, I wondered, had happened to the air conditioned Land Rovers?

Once through customs, with our baggage collected, a group of exactly one hundred intrepid explorers were gathered together by *Safari*'s representatives. We were taken to our transport. The trucks were 4-ton *Bedford RLs*. The yellow paint had been newly and hurriedly applied and underneath it, KHAKI shone through. They were ex-Army, sold off and driven up from Aden when the British withdrew from its former protectorate. We loaded our cases into the trailers, then climbed aboard the trucks, using a conveniently placed knotted rope and found the 'comfort' provided by twelve old, two-person, bus seats that had been screwed down in the back of the trucks, six seats to each side. I was less than impressed.

We drove a short way along the coast from Tangier and stopped at a small and very basic resort centre, situated just behind a beautiful sweep of deserted, sandy beach. The centre was recently built and comprised administrative, shower and toilet blocks, store buildings and a café. We were issued with a pup tent, sufficient to take two people, and sleeping mats. After telling us when and where a briefing would be held, the *Safari* staff indicated the areas where the pup tents were to be erected and left us to it.

The *Safari* team had programmed the first three days to be spent at the centre, with a couple of trips to places of interest. It was sufficient time for some acclimatisation to take place and for the sillier sunbathers to be stopped from being burned to a crisp. The initial briefing had asked the hundred holiday makers to form themselves into four groups of twenty-five and then to decide which of the four *Safari* drivers they wanted to take them on their tour of Morocco. There was no set itinerary. Where we went varied according to a group's wishes. Sue and I were doubtful about this *laissez-faire* approach, but the system worked and within the three days, people gravitated towards one another and the groups formed naturally.

Our group set off heading south west towards Rabat. That was the start of a series of interesting adventures that occurred more or less daily. On day one the Bedford grumbled its way along for hours, then gave out on a main road, in the middle of nowhere, with night approaching. The group established camp next to a dry riverbed and went supperless, while Sue (French speaker) and I hitched a lift to the nearest town to contact *Safari*'s base for help – and to have a decent meal. Help came in the early hours of the morning. Whatever the mechanic did, it worked and the Bedford chugged along happily for the rest of the holiday.

The group had decided to meander down Morocco's western coastline and return by an inland route through the desert. Nothing was set in stone and we varied our route as the mood took us. We camped all the way, rarely, if ever, bothering with tents. Nights under the stars were wonderful, a first for everyone other than the driver and myself. Overnight venues varied from a comfortable camp site in a resort town, to a *Beau Geste*, desert fort that had been largely demolished (blown up) by the French Foreign Legion before they left the country.

As we headed further south, the food that we were able to buy became less westernised and the real rough camping and cross desert motoring began. In the desert the old Bedford proved its worth. It was a high bodied vehicle, with a metal frame constructed over its back to support a canvas roof and sides. Sitting on our bus seats, we were well above the wind-blown sand that would have made travelling in a lower vehicle unpleasant, unless reliance was placed entirely on air conditioning. The canvas roof provided good shade that could be augmented by side panels as necessary.

There was one heart-stopping moment when the truck bogged in soft sand above a remote sea estuary. We were short of water and because mobile phones were still a long way off, we had no means of letting anyone know where we were. It took a lot of digging, the use of sand decking and much cajoling of the less willing, to extricate ourselves.

We harboured in open desert beyond the Berber town of Tan Tan. There a guard had to be mounted by the men of the group, because on a previous trip a local guard had been murdered by tribesmen. The boys were all heroes in the morning, as they sought to impress the girls with stories of moving shadows and strange noises in the night.

In the same Berber country we bought half of a newly slaughtered sheep. The butcher did not joint it for us and Sue, as the only married woman in the group, was appointed chef. In the circumstances, her sandy tagine was a triumph.

Most remarkable of all on this part of the journey, our driver knew of an ancient baths, built in a tiny oasis. We were fairly grubby by the time we got there and were delighted that it had a runnel that ran all the way around the outside of the central pool. The pool was for bathing; the runnel was designed for washing. A Scottish girl shampooed her very long hair while sitting next to me and asked if I would comb it out. I passed the comb just once through her hair. Its teeth gathered tens of tiny red worms! Her scream must have been heard in Casablanca. We all spent ages getting rid of the things.

Our way back to the northern coast was via Marrakesh and the Atlas mountains. Marrakesh was memorable, firstly, because we stayed in an Arab hotel, which provided the only bed of the entire holiday and, secondly, for its history, architecture and souq. The souq was a warren of narrow passageways, crammed with tiny shops that sold absolutely everything from carpets and copperware, to spices and spare parts for cars. Camels were sold in the main square, alongside dentistry, with patients squatting nervously on a mat, while the 'dentist' operated with a pair of electrical plyers! The colours, smells, clamour and press of people was exhilarating.

The final leg, over the Atlas Mountains, was undoubtedly the hairiest drive I have ever endured. It was over unmade roads, with vertiginous drops and oncoming drivers intent on suicide. The scenery was absolutely magnificent – breath-taking in fact. We spent our last two nights in the *Safari* resort, where I managed to dislocate my right elbow playing football on the beach. Luckily, I was able to reset it myself, but it had swollen to look like elephantiasis by the time we reached the UK. It took time and a lot of physiotherapy to mend.

Warminster was comparatively tame after our Moroccan adventure, but our lives were enlivened by additions to the family – not babies, but kittens. Sue had always been a 'cat person'. The lady in the quarter opposite to us suddenly turned up with a gorgeous, long-haired, ginger kitten. She had got it from a local farm and there were two kittens left. She said that they were to meet a watery end if someone did not adopt them quickly. That was enough for Sue. We drove immediately to St Michael's Farm and the kittens were ours in seconds. We had two ginger toms. One was all over ginger. The other had a white chest and white socks on his feet. We had to find names for them. As they came from St Michael's Farm, it was not too difficult. The one with white was 'Marks' – the other became 'Spencer'. They were characters and, much later, became 'by Royal appointment'.

Well into my second year at the Skill at Arms Division, I was told that in the autumn I was to return to the battalion, which by then had moved to Paderborn, in northern Germany. Additionally, I was told that I was being considered as the next adjutant, but that it would depend on the approval of the incoming commanding officer, Jonathan Hall-Tipping. We had never met as his background was with our 1st Battalion. At the time Jonathan was serving in the MOD in London and we arranged to meet for lunch. We took an immediate liking to each other and my next appointment, as adjutant, was quickly confirmed.

I have pointed out previously that adjutant is a demanding appointment, albeit one that is largely deskbound. To improve the staff skills of captains, the Army had recently introduced a course of about ten weeks, held at Warminster, organised by the Junior Division of the Staff College (JDSC). It covered every sort of writing that might be encountered in a headquarters and it was very intensive. Most nights saw the burning of midnight oil. However, officers emerged much the better for it and most were confident of their ability to hold down a grade 3 (captain's) staff appointment. It would be ideal preparation for being adjutant and, as I was serving at Warminster at the time, it was entirely convenient for me to attend the course before Sue and I left for Germany.

Rarely is anything straightforward. I made contact with the battalion to learn that when I was due to arrive with them, they would be the nominated *Spearhead* battalion in Germany. During my tour in Warminster, the IRA had begun its campaign in Northern Ireland and terrorist activity was already on a sharp upward curve, requiring the Army to send reinforcements to the Province to combat it. The *Spearhead* battalion was placed on short notice to move to Northern Ireland should the situation worsen and the smart money was on the Pompadours being deployed there during the time they held this commitment. It made little sense for Sue to give up her job in Frome, go to Paderborn, where she would not have one, while I would probably be in Northern Ireland for several months. We quickly decided that she should stay in England. However, there was another problem to overcome. Once I had been posted to Germany, we would have to give up our quarter, find somewhere else for her to live and hire it privately. The place we found was formerly a farm worker's cottage in nearby Old Dilton. It would prove to be as cold, if not colder, than Elm Hill. In my absence, the cats slept with her to keep her warm!

I completed the JDSC course and had leave due to me before I was required in Paderborn. We took it with a fellow syndicate officer, Jimmy McSheehy, and his girlfriend, Angie. We decided to drive independently across France to Fréjus, on the Cote d'Azur, and join up for a beach holiday. We had a great time and it was just the fillip needed before Sue and I had to say goodbye and I set off to re-join the Pompadours.

CHAPTER EIGHT

THE PADERBORN/NORTHERN IRELAND SHUTTLE

THE JOURNEY TO West Germany was by commercial airliner from Luton Airport to RAF Gutersloh. The airport at Luton was in its early stages of development, so had little to commend it and the air movements staff still insisted on interminable, RAF waiting times, despite the fact that a civil airline held the contract for UK – British Army of the Rhine (BAOR) movement. On leaving the Gutersloh arrivals hall, I thought that as I was the incoming adjutant, I might at least have a Land Rover waiting for me – but no such luxury was forthcoming.

On arrival at Alanbrooke Barracks, Paderborn, home to the 3rd Battalion, I made my way to the officers' mess and settled into my bedroom. The battalion was deployed on exercise locally, so my first evening back with the Pompadours was distinctly downbeat, spent in an all but empty mess. The following morning I went straight to the battalion headquarters building to be told that I was to join the exercise as soon as possible.

During my time at Warminster the battalion had become a mechanised unit, equipped with *Mark II AFV 432s* (tracked, armoured fighting vehicles). I knew very little about the vehicle, or armoured warfare, so was happy to have an early chance to get acquainted. In the field, battalion headquarters was tactically deployed under camouflage netting. Everyone, other than me, carried a weapon, wore camouflage cream, steel helmets and full webbing equipment. I was not concerned to be the odd man out as, up to then, I had had no chance to draw any of that kit. I opened the door of the command APC (armoured personnel carrier) only to be roundly cursed from inside and told to close the fucking door. I recognised the commanding officer's voice. It was the same officer, Lieutenant Colonel Burch, who had taken command just before I left the battalion in Aldershot.

While I was at Warminster I had maintained regular touch with friends still serving with the battalion. They were unanimous in the view that the commanding officer was an extremely able soldier, but his extreme, authoritarian style all too often bordered maltreatment. Several officers had been sacked, or moved on, and the rank and file were profoundly discontent. After I assumed the adjutant's appointment, I found that over eighty soldiers had applied to purchase their discharge from the Army, such was their unhappiness. All of that was directly connected to the deeply unpleasant way the commanding officer chose to exercise his authority. I was to experience it personally in the days and weeks that followed.

The exercise concluded that evening and the following morning I began the handover/takeover from my predecessor, Hugh Lambert. I had known Hugh since my first day in the battalion at Ballykinler and he was a particularly industrious officer. He was now due a posting and I would become the fourth adjutant in less than two and a half years – not a happy record for any commanding officer.

I took over at close of play on a Friday evening. After dinner, in the mess, I was listening to the evening news when it was announced that another unit was to be sent immediately to reinforce the Northern Ireland garrison. The news went around the mess like wildfire and I was about to go to battalion headquarters to update work on the battalion's fly out plans, when a mess waiter said that the commanding officer wanted to speak to me on the telephone. I picked up the phone and said, 'Yes, sir.' With no preamble, he aggressively demanded, 'What's my fly out strength?' I said that I was about to go to battalion headquarters to confirm that and update any other detail that would be needed. He swore at me and told me that he would never have selected me to be adjutant. Moreover, I had to improve if I was to remain in the appointment. 'Ring me back when you've got some idea of what you're doing,' was his parting shot in a conversation that lasted less than twenty seconds. I did ring him back, but within an hour or so, HQ BAOR decided that the *Spearhead* battalion was to remain just that and another unit was to be sent. Apparently the move was not as urgent as had been announced on the news. The reinforcement would be made over a week or so and it was thought better to hold the *Spearhead* battalion, on its short notice, in case a really pressing emergency occurred in the meantime, which required a further and faster buttressing of the garrison in the Province.

In bed that night I consoled myself that I would only have six weeks to endure before Jonathan Hall-Tipping arrived to take command. It looked like being a long month and a half and so it turned out to be. Every day the atmosphere in the office was tense. Mine was an outer office to the commanding officer's and as I was his personal staff officer there was no way that we could avoid one another. Our mutual dislike was evident, but it was mitigated somewhat because virtually everyone who went in to his office was on the wrong end of some sort of unpleasantness. I was definitely not alone.

The dam broke one morning when the bell went in my office, indicating that the commanding officer wanted me. I went into his office and closed the door. He was writing at his desk.

Without looking at me he said, 'Take this lot away,' and shoved his 'out' tray violently. It flew off his desk and crashed to the floor scattering the files.

I swallowed. 'I am the adjutant, sir. I deserve better than that.'

I turned, left his office, and then sat shaking at my own desk. I thought I was going to be sacked.

'What's the matter?' Tony Downes, my assistant adjutant, sat opposite me; he had been commissioned from the ranks and was formerly a 1st Battalion man, so I had not met him before arriving in Paderborn. He was the steadiest of hands and helped me regain my composure. The bell went again. The 'out' tray was back on the commanding officer's desk, with its files in place. 'They can go,' was all he said. I picked up the tray in silence and went back to my own work. From that time until he left, a sort of armed truce existed between us and further major outbursts of nastiness were avoided.

When his final day in command came, it was a departure from tradition. 'Final days' are big events, because they are a commanding officer's last ever day at regimental duty. He, or she, then moves to senior staff, or other appointments, and will never again serve in a unit. To mark that major milestone, a departing commanding officer is pulled out of barracks by members of the officers' and sergeants' messes, generally in an open-topped vehicle of some sort, with the route lined by the unit's soldiers. However, Lieutenant Colonel Burch said he did not want that ritual to take place. Instead, he wished to address the assembled battalion in the gymnasium and then leave by his staff car. I made the necessary arrangements with the RSM and when the time came, the CO and I drove to the gymnasium.

As we approached the building, he turned to me and said, 'You don't think they'll boo me, do you?' I said I thought that unlikely.

He spoke well, but was heard in silence. We left and on the way to the barracks' gate he looked utterly miserable. I got out of the car at the gate and saluted as he was driven away. I determined that should I ever get the chance to command, I was not going to leave like that.

Lieutenant Colonel Jonathan Hall-Tipping arrived very shortly afterwards. The first thing he did was to smile, talk and take an interest in everyone he met. What a relief that was. The atmosphere lightened almost instantly. He was a 'people person' and he needed to be. I had gathered that from our first meeting at lunch in Whitehall and the view was reinforced when Sue let me know that unexpectedly one evening, her cottage door in Old Dilton was knocked and Jonathan and his wife, Joy, had travelled to meet her and to take her out to dinner. The kindness was typical of them both.

Jonathan Hall-Tipping had inherited a remarkably efficient, well trained, APC battalion, but its morale was at rock bottom. That needed to change and change quickly because the Pompadours had been warned of a four and a half month, emergency tour in Northern Ireland to begin in late March 1972, which was just over four months away. The armoured vehicles and our preparedness for high intensity warfare in northwest Europe would be put to one side. In their place would come internal security training, leading to an operational deployment in the United Kingdom.

The new focus and the excitement promised by an operational tour greatly assisted Lieutenant Colonel Hall-Tipping in his efforts to lighten the battalion's corporate mood and re-build its team spirit. He was helped further by the strong team he had around him. Major Leon Paul was an experienced officer, trusted confidante and effective second-in-command. Two of the company commanders were particularly able, Majors Alastair Veitch and Bill Dodd. They were supplemented by a clutch of talented captains, working in key appointments in battalion headquarters; Brian Davenport (Operations Officer), Julian Browne (Intelligence Officer) and David Norbury (Regimental Signals Officer). Jonathan immediately recognised the strength of that team and quickly communicated his appreciation of their worth. The Christmas period was sensibly used to regenerate a sense of fun that had been absent for far too long. There seemed to be more parties than ever and well before the New Year, the Pompadours were back on an even keel.

Undoubtedly, my best Christmas present was to be re-united with Sue. Just before the turn of the year, we took over our first quarter in Germany – the cats having been lodged temporarily with Sue's aunt. No 10 Rudolphiweg was very different from 82 Elm Hill. It was a semi-detached, three bedroomed house, without a garage and only a small rear garden, but it was insulated, double glazed, with window shutters, and it had cellars that housed two, cantankerous, coke-fired boilers (a big one for winter and a small one for summer). And the house was warm!

Training began immediately in the New Year. Specialist advisers, who would develop centralised training programmes for the internal security role, were in the process of being brought together, so commanding officers still had to devise their own training in readiness for the task ahead of them. Our area of responsibility in Belfast was to centre on the Falls, the staunchest of all republican communities and the hardest of 'hard' areas. We had drawn the tough one. To help us orientate to our new role, Alanbrooke Barracks was re-purposed. Its roads suddenly became Albert Street, the Falls Road, Milford Street and Northumberland Street. Buildings got new names to reflect those we would get to know in Belfast. Individual training (fitness, shooting, learning the local 'wanted' lists and geography from photographs and maps, and ingraining the rules of engagement), led to company exercises, which culminated in a three-day battalion exercise. The 'enemy' was provided by the Recce Platoon and the mechanics of the Royal Electrical and Mechanical Engineers (REME), who were attached to us and are an integral part of any armoured unit. They provided daily shooting incidents, bombings and some extremely realistic rioting. When all that was completed, we had done all we could to prepare ourselves for the coming ordeal.

All of the battalion's 'patch' in Belfast was in hard areas. Battalion headquarters was located in Hastings Street, just off the Falls Road and adjacent to a RUC (Royal Ulster Constabulary) station. In the centre of 'A' Company's area was the Divis flats – a monolithic structure of some nineteen floors that would not have been out of place in Stalinist Russia; 'B' Company was in the Clonard area; and 'C' Company oversaw the Lower Falls. All of those were strongly Catholic and republican. However, the so-called 'Peaceline' ran through our area and Support Company policed it and the diehard loyalist, protestant area that took in part of the Shankill. The potential for trouble was enormous and it arrived quickly.

We took over from 1st Battalion, Gloucestershire Regiment. Near the end of their tour, the government had tried a peace initiative and the Army had adopted a low profile to support it. We had only a few days of relative quiet before the IRA instigated some severe rioting, mainly in the Divis flats area. It lasted intermittently for most of the first half of April. It was literally a baptism of fire, particularly for 'A' Company.

My contribution to operations throughout the tour was to be a principal watchkeeper, manning the forward radio net from battalion headquarters to the companies. It was outrageously busy. Messages, reports and orders to adjust deployments were constant for much of every day and everything had to be logged in case the logs were required as evidence – those logs are still providing vital evidence, nearly five decades later. Next to the forward link watchkeeper was another officer watchkeeper – his job was to man the rear radio link to brigade headquarters, summarising the events that he could hear coming in over the forward link, and passing on any instructions from the brigade staff. The work was not at all dangerous, but by the end of a long shift, the mental effort required of everyone engaged in the operations room left us drained.

I did go out with foot patrols from time to time, because it was important that the soldiers, who patrolled daily, saw officers from battalion headquarters with them on the ground. However, they were practiced teams and they had to alter their established routines to accommodate me, so it was not something to be done without some thought.

Researching the battalion's newsletters and magazines of the time, I found that over 1,000 rounds were fired at the battalion during our initial weeks in Belfast. We were lucky to survive them without fatalities. But this streak of good luck that extended from our tour in Aden, was not to last. On the 16 April, Second Lieutenant Nick Hull was killed, sitting in the commander's seat of a wheeled APC, a 'Pig', while it was mobile. Nick was a very likeable young officer. His platoon had won an inter-platoon competition shortly before we had left Paderborn; he was a growing talent and would be greatly missed.

Nothing was stable for very long. Our boundaries were changed quite often reflecting the ebb and flow of violence. At various times during our tour, we had charge of the city centre and helped introduce a pedestrian only area there;

we oversaw the Sandy Row (loyalist); and the Markets (republican). From time to time, we had up to five additional companies, attached to us from other units, to help us control these expanded areas. It made us, by far, the biggest unit serving in the Province.

Shootings and bombings continued and, tragically, on 11 May, Private John Ballard was killed on foot patrol in the Lower Falls. He was in 'C' Company, as was his brother, who was part of another patrol that went to his aid. Finding his brother murdered was dreadfully traumatic. The commanding officer was very concerned about the effect that would have on the surviving brother and his parents. I was asked to take my R&R (Rest and Recuperation) leave early and to go to Grimsby for the funeral, where I did what I could to explain to Mr and Mrs Ballard the hurt felt in the battalion, the situation in Belfast and the conditions in which their sons had served there. How much good I did I am not sure, but the parents appreciated the efforts that the Regiment made to support and comfort them.

A big plus about taking early R&R in England was that Sue was staying with her parents and I was able to spend a couple of days with her. That happy time was over far too quickly and soon after my return to Belfast, more rioting erupted in the Divis flats area that presented me with a possible sniping opportunity. At the beginning of the tour I had made a thorough examination of the former mill in which battalion headquarters was accommodated. It had an attic space, with a pitched roof and a couple of the roof tiles had slipped. I got a chair and looked out of the gap that had been created. I had an excellent view of the upper half of one aspect of the Divis flats, about 500–600m away. It was a place from which IRA gunmen had been known to fire. I decided to develop this vantage point as a sniper hide. I had a tall stool made that elevated me sufficiently to be able to fire through the gap in the roof if I moved the loose tiles a little further apart. With rioting in progress, I went to my roof hide. I could not see the riot, but a gunman had been reported in the Divis, so there was a chance that he might fire from the part of the complex that I could see. It was already dark, wet and windy. Normally, gunfire and wet windy conditions would be enough to convince rioters that the TV, or pub, were better options, but not this night – violence continued.

Suddenly, I saw muzzle flashes from one of the darkened windows, high in the flats. I identified the window, but could see no more detail. Then there were two more muzzle flashes. My predicament was that I could not see the gunman, therefore I could not positively identify the target, but I knew exactly from where he was firing. If I decided not to fire, the gunman might well go on to kill or injure one or more soldiers trying to control the riot below him. I was confident of being able to hit the window, even on a dark, windy night, so I carefully fired one round and waited ... and waited. After some thirty minutes, no more firing came from the window, or anywhere else that I could see. I reported my action because soldiers, rightly, had to account for every round they fired in Northern Ireland. Searches were made in the flats the following morning, but without conclusive result.

Even in those dire days, we were able to cling to some vestiges of normality. The battalion had a remarkably talented football team at the time. I was part of the squad, but no longer the first choice goalkeeper (Private Northrop was not only first choice for the battalion, he was the Army's 'keeper). We did our best to keep our fitness levels up in Belfast, because midway through the tour, the squad had to return to Paderborn to play in the Infantry Cup final against 1st Battalion, The Queen's Own Highlanders. For a long while it was a closely contested game, before we got on top and won by a convincing 3–0 margin. The result might not have meant anything to anyone outside the Pompadour family, but it gave the battalion a significant lift.

Violence in Northern Ireland continued throughout May, but at the end of the month the Official IRA declared a ceasefire. They were not joined by the Provisional IRA until nearly a month later. Nevertheless the Army was ordered to lower its profile once again and that was the opportunity for protestant loyalist gangs to seize the initiative. The Ulster Defence Association and the 'Tartan gangs' grew in size amazingly quickly and they made attempts to establish no-go areas in some protestant communities. The battalion had one particularly close call when tens of thousands of loyalists marched from their strongholds, intent on invading the republican Lower Falls. They were stopped on the *Peaceline* and turned around by the commanding officer, RSM and only a single platoon (a maximum of thirty men). Had they stormed into the republican areas, the mayhem would have been unimaginable.

On 9 July, the Provisional IRA broke its ceasefire that for them had lasted barely a fortnight. Gradually, the Army's profile was raised in response. Double tragedy struck on 13 July, when first, Lance Corporal Martin Rooney and then Corporal Kenneth Mogg, were killed in similar circumstances, murdered by armour piercing rounds that penetrated the 'Pig' APCs in which they were travelling.

I knew Corporal Mogg particularly well. He had been in my company in Aden and at the School of Infantry while I was in Warminster. Arrangements for the notification of relatives was my responsibility. The protocol was that news to next of kin living in the UK was conveyed, in person, by a member of the local civilian police. Corporal Mogg was killed late at night. From his personal record I found that his wife was living in England. I got the telephone number of the local police station, spoke to the duty sergeant at about two o'clock and made absolutely clear to him that under no circumstances was Mrs Mogg to be told before the morning (I think seven o'clock was the earliest permitted time). That restriction was designed to ensure that the support of relatives and friends would be much more readily forthcoming than in the middle of the night. Additionally, regimental representatives, already alerted, would have the chance to get their support to her with a minimum of delay. Thirty minutes after speaking to the police sergeant, a totally distraught, grief stricken Mrs Mogg tried to speak to me on the telephone, asking me to tell her that it was not true and her Ken was alive. I have never felt so inadequate. I could do nothing to stem her anguish, except to move heaven and earth to get support to her. I was outraged at the appalling cockup made by the police, but my protests were waived away. I was simply told that there had been a genuine misunderstanding.

Lance Corporal Rooney was a Catholic and a southern Irishman. We always seemed to have a number of 'Micks' in the battalion and they normally served with distinction. The Garda were a great deal more sympathetic and efficient than the British police and the subsequent telephone call I had with a family member was measured and dignified. The family wanted Lance Corporal Rooney's body repatriated to the Irish Republic. We asked if our padre could accompany the body to the family home and this request was readily agreed. Captain Reverend Jim Symonds was a remarkable man. He remains an Anglican priest and had taken holy orders in his mid-forties. He was a man of the world and a really good football referee – both admirable ingredients for an Army padre. Not only did he accompany the body to the family, but after only a few hours, he was asked to help officiate at the funeral. I am still staggered to think that an officer and protestant priest, prayed over the body of a Catholic soldier of the British Army, at his funeral, in the Republic, at the height of the troubles. It would be difficult to make it up, but it spoke volumes of 'Jim the Hymn' and the generosity of the Rooney family.

It was at about the same time that I was required to operate as a sniper. A sangar (sand-bagged sentry position) outside Support Company's headquarters had been fired on from a school at a junction with the Falls Road. It had happened on two occasions between 21.00 and midnight on successive Thursdays. It could be a pattern was emerging. A derelict house at the end of a terrace overlooked the school and would provide a place for a sniper to be positioned. I was to occupy a first floor room of the house with a No 2, Colour Sergeant John Fisk, and hope that the terrorists would try their luck while we were there. I was the sniper. Colour Sergeant Fisk was there to provide my close protection. I was armed with a L42, bolt action sniper rifle and Browning 9mm automatic pistol. He had a standard issue 7.62mm SLR (self-loading rifle). Our chosen hide was just across a main road from the school, so that our likely target area would be only 40 metres from us.

We were inserted in the small hours of the morning, using a ladder from the back of a Pig to get us onto a tall brick wall, then across the flat roof of a substantial shed and into the derelict house, via a rear window – the ground floor doors and windows were all boarded up. The Pig and its crew had left even before we were across the flat roof. We were on our own, however, back-up would always be only a matter of a couple of minutes away. What was unsettling was the fact that we did not know if we had been observed entering and we could not see the street directly below our window. It was very possible that if we had been seen, a bomb could be placed beneath us without our knowing.

As daylight broke we noticed that a brick had been loosened in the end wall of the room that we had occupied. We were alarmed. We removed the brick carefully and it confirmed our worst suspicion. We were not the only ones to recognise a good fire position – we were in a hide used by the IRA to overlook the Falls Road. We waited throughout the daylight hours, taking turns to observe the school, but there was nothing untoward in the scene below us. Towards evening a man climbed onto the roof of the shed adjoining the rear window through which we had entered the house. He poked around amongst some timber stored there, but made no attempt to look into our building. We decided that we were safe enough.

Darkness fell and street lamps threw sufficient light into the school's playground for us to be sure of a good sighting if anyone approached its low, perimeter wall and railings that adjoined the street. The area beyond that illuminated patch was in impenetrable shadow and provided an excellent covered approach to the wall until the last fifteen yards across the open playground. If anyone appeared, I would not have long to make a decision to fire.

At one point we thought there might have been movement in the shadow, but it came to nothing. Dawn was not far away and we had done our stint. Our Pig made several intermittent passes up and down the road making no attempt to extract us, before it pulled in beside the tall brick wall. We were across the shed's roof, down the ladder in a trice and into the Pig. Its rear doors closed and we were away. It had probably been stationary for about ten to fifteen seconds. It had been an unproductive stake out, sometimes trying on the nerves, but for whatever reason, the school was not used as a firing point again by the IRA.

Throughout July the security situation across the Province got progressively worse. That was particularly true in Londonderry, where the Bogside and Creggan areas were heavily barricaded and no-go areas existed briefly. Trouble in Belfast reached a dramatic crescendo on Friday, 23 July, Bloody Friday. Over twenty bombs, mainly car bombs, detonated in the city centre in less than an hour and a half, with the bulk of them exploding in a concentrated thirty-minute period. The Provisional IRA was responsible and claimed to have given the security forces adequate warning, but that was certainly not the case. Many of their warnings were hoaxes, adding further confusion to the mayhem that they had already created. The tragic result was nine deaths and 130 injured, but there was a public backlash against the IRA. In keeping with other units in Northern Ireland, the battalion stepped up its operations significantly; raided many, many houses and made a significant number of arrests. On 31 July, Operation Motorman was mounted simultaneously across the Province. In Londonderry the barricades were removed. House raids and arrests continued to be made in Belfast and the battalion played a full part in that. Some semblance of order was restored and we handed over to our 2nd Battalion, in early August, to return to Paderborn.

Even before we had left Belfast an impressively thick document had arrived from Headquarters 4th Division, which was the major formation that commanded us in Germany. The document was the detailed instruction for a formation training exercise (FTX) that was to take place in late autumn, in the Hartz Mountains, quite close to the border with the DDR. The exercise instruction was a not so gentle reminder that we would have to get back into the mechanised role very quickly on our return to Germany. However, that could wait – first we were to take three weeks block leave, when virtually the entire battalion was to go on holiday, all at the same time. For Sue and me that meant an interesting tour of Yugoslavia, which was the only communist state that I was allowed to visit because its head of state, President Tito, was fiercely independent of the Soviet Union and kept his country genuinely non-aligned. It was considered a safe place for western soldiers to visit.

As soon as we returned from block leave, the APCs were loaded onto flat-bed railway trucks to be taken to Soltau for two weeks work-up training with our associated tank squadron, artillery battery and engineer troop. From there we moved directly to the Hartz Mountains for the FTX that lasted three weeks. It was testing and a huge contrast to operations in Northern Ireland.

The threat from the Warsaw Pact forces throughout the Cold War period was so great that every autumn, once the harvest had been gathered, the various armies present in West Germany were allowed to exercise over normal countryside and rumble through any small towns and villages that were in their way. Because the areas involved were many times larger than even the biggest military training area, divisions (formations of around 30,000 soldiers) and, occasionally, a corps (in excess of 100,000 soldiers), could be realistically manoeuvred. Compensation was paid for any damage, but the disruption to the civilian populace was enormous. They understood the rationale for the exercises, but we were not popular.

After the FTX, the battalion had to undergo its fitness for role (FFR) inspection. Our brigade commander was Brigadier Dickie Lawson, an extremely talented tank officer and, later, a senior general and member of the Army Board. He thought that the Pompadours had had a pretty tough year, so readily agreed that if we jumped through all the inspection hoops, we could have a bit of fun. We wrote an outrageous scenario that required the battalion to impress a visiting 'Arab' general and in the process we would go through all the necessary FFR inspections. On the day, the brigadier surprised us all by turning up in a shemagh headdress. It was his turn to be surprised when I produced a camel, hired from a local circus. To his great credit, he later rode the beast to inspect a parade of armoured vehicles, part of the programme to impress.

Christmas festivities came next and then the New Year was upon us. 1973 transpired to be a virtual mirror image of its predecessor. Again the Pompadours had been warned for an emergency tour in Northern Ireland. The dates of the deployment almost exactly matched those in 1972, but this time our destination was Londonderry – a case of out of the frying pan and into the fire.

Sue and I had another priority. She was pregnant and her due date was fast approaching. I learned how to puff and pant with her, but that was about my sole contribution, other than to drive her to the British Military Hospital (BMH) Rhinteln, about 60 miles from Paderborn.

The BMH was a former Wehrmacht hospital. Extraordinarily, it was built in the shape of a swastika. It was staffed by members of the Royal Army Medical Corps and the Queen Alexander's Royal Army Nursing Corps. They were proficient, but Sue found that many lacked empathy with wives. Some of that may have resulted from the fact that wives did not have their own identity. Sue's bed card read 'Wife of Captain C. Groves' followed by my Army number. That loss of personal identity did not pertain to the hospital alone. Sue could not get a book from the Army library, or buy her groceries by cheque from the NAAFI, without quoting my name and number. It was de-humanising, but we somehow accepted it.

When, at last, she was taken in to hospital to have her baby, it was to be a long job. Poor love, she was in labour for over thirty hours, getting weaker and weaker all the time. For the last twelve hours or so, she was so sedated that she was incoherent and not properly conscious. Finally, I was asked to sign for a caesarean birth (she was in no fit state to sign, or to make a decision). As she was wheeled to the delivery room, she suddenly sat upright on the trolley, looked straight at me and said, very clearly, 'I hope we like it!'

'Bit late now, darling,' I thought.

Nicholas appeared after a forceps delivery, not a caesarean section, but I was not allowed to be present. However, I was permitted to see Sue immediately after the birth, but she was not really aware of me, or anything else. A female staff nurse took me to see my son in the nearby nursery.

Soldiers and their wives are young and healthy. What they do well – and often – is to produce offspring. The nursery had at least thirty babies lying in canvas cots. The staff nurse went around reading the cards on the cots, before picking up a tiny bundle. She gave him to me to hold. It is difficult to describe just how intense and immediate the connection was between me and the minute human being I was holding. I did not expect a sensation so passionate and so strong. He was mine – ours. Sue had got her wish – I did so much more than merely 'like it'.

There were a lot of babies born after the tour in Belfast, but Nick was the first – a fact that did not escape the soldiers' notice. On my way to see Sue at Rhinteln, I was stopped by a West German police car. I was perplexed because I was certain I had done nothing wrong. The policeman waved me out of my car and indicated my rear number plate. At that time, British military personnel had registration numbers for their private vehicles that differed from German ones. Germans had black letters and numbers on a white background. Ours were the reverse; white on black; with letters, numbers and a final 'B'. I forget the detail of my number plate, but it would have been something like 'FHD 363 B'; now it read 'R AND R 1'. It took some explaining, but the copper saw the funny side and, very kindly provided an impromptu 'blue light' escort most of the way to the hospital.

On her return home, Sue and I had just four weeks together with Nick, before I was on a plane to Belfast's Aldergrove airport and the start of more than four months' separation. That was not easy for anyone, but doubly difficult for a wife with such a new, first baby.

In the meantime, Alanbrooke Barracks had undergone a now familiar transformation. Sandbag sentry emplacements and unloading bays appeared and the road names changed again, this time to Lislane Drive, Creggan Heights, Fanad Drive, Brook Park and other Londonderry favourites. The soldiers poured over maps showing our new areas of responsibility, all of them avidly republican, which included the Creggan, Rosemount and Glen Owen estates. In addition, we patrolled a rural enclave and its border with the Irish Republic. The soldiers studied fresh wanted lists, but the Yellow Card (the rules of engagement), was already committed to memory for the vast majority. We were treading a well-known route.

However, in just twelve months, a lot had changed on the pre-training scene. Now the nearby Sennelager Training Area had an excellent urban close quarter battle range for us to use and a realistic 'Tin City' that replicated the urban areas of Belfast and Londonderry, where patrol, anti-riot and vehicle checkpoint techniques could be practised. Additionally, search team training had become much more specialised and the CO and company commanders were given television technique training, because the Army had learned that the international media would constantly be on the lookout for stories and quotes.

The battalion took over its area from 1st Battalion, Grenadier Guards in March. The weeks leading up to the handover had been relatively quiet, but just as in the previous year, we got away to a hot start, with four days of rioting in the Creggan estate. Gunmen made their appearance in April and we were to suffer our first and only fatality on the 27th of the month, when Private Anthony Goodfellow was murdered by an IRA sniper, while manning a vehicle checkpoint.

The heightened level of violence was maintained during May and June as the IRA terrorists widened their repertoire of attack techniques to include land mines, later to become known as improvised explosive devices (IEDs), a rocket attack against an APC and a botched mortar attack against Creggan Camp, where most of the battalion were located. Those months also brought a new requirement – local and assembly elections were held and polling stations immediately became high value targets for terrorists. The polling stations needed guarding and many house searches were mounted with a view to making pre-emptive arrests to nip terrorist attack plans in the bud.

My part in all this was virtually unchanged from the Belfast tour. I was still a principal watchkeeper in the battalion's operations room, but I was not needed as a sniper. The Creggan estate, where the bulk of terrorist activity took place, was a relatively modern development. There were very few derelict buildings and the street layout was wide and generous, making covert insertion difficult. The chances of not being seen were slim. Finding suitable hides was near impossible and any that were identified had to be tied to specific intelligence that indicated that some worthwhile target might appear. All too difficult, with little chance of success, so my L42 rifle was parked.

I was fortunate to see a few days of action. 'A' Company's commander, Major Kerry Woodrow, was due to take his four days of R&R leave. His second-in-command would normally have commanded in his absence, but was suffering a crisis of confidence that later caused him to transfer to a logistic service. The CO told me to step in.

'A' Company's base was the Blighs Lane post, a former factory building, overlooked by nearby housing. That was particularly dangerous because the potential firing points provided by the houses, posed a constant threat from gunmen. Rioting was rife and every day that I was in command, the company deployed to contain it. On one occasion I was directing operations in a residential street and my attention was firmly on the rioters in front of me, many of them children and therefore, doubly difficult to contain without causing an outcry. There was a flurry of movement immediately behind me and a woman of about thirty lay moaning at my feet.

'She was going smash your head with the edge a dustbin lid.' It was Company Sergeant Major Roy Brunning who spoke.

An old fashioned, round, iron, dustbin lid lay next to the woman. I nodded my thanks and got on with sorting out the riot.

Until very recently I was under the impression that Roy Brunning had saved my bacon that day, however that was not the case. Earlier this year (2020) I got an unexpected telephone call from a Mr Wilkins. I had known 'Wilkie' from Berlin days – he was another footballer. His telephone call came about because he had just been informed by the Ministry of Defence that he was to be investigated for his actions in Londonderry. He was a corporal then and I was probably in command of 'A' Company at the time to be investigated. During our conversation it turned out that it was not Sergeant Major Brunning who had saved me – it was Corporal Wilkins. He had seen my would-be assailant approach from behind me and had intervened in the nick of time. He had re-joined his section without saying a word. However, it was not this incident that was being investigated and Mr Wilkins wanted to know if I could shed any light on the one that was under the spotlight. He could not relate to it and neither was I able to help. It later transpired to be a case of mistaken identity. Mr Wilkins was exonerated, but while the investigation hung over his head it caused a good man, now in his seventies, very considerable and avoidable anxiety.

On a lighter note, the battalion football team was still in action and another final of the Infantry Cup had to be played back in Paderborn, this time against 1st Battalion, King's Own Border Regiment. We trained as a squad a couple of times a week in Londonderry, initially in a disused factory in a protestant area of the city. After a training session, we were returning to Creggan Camp in two Land Rovers and had reached the republican side of the River Foyle, when we were fired on by a gunman. Both vehicles screeched to a halt and we all 'debussed' and ran towards the gunman's position. He must have been amazed. Instead of uniformed soldiers in combat kit, flak jackets and boots, we were armed, but wore tracksuits and trainers and were running a great deal faster than any soldier in standard uniform. However, as was virtually always the case, the getaway route was uppermost in the gunman's mind and he made good his escape through a labyrinth of alleyways and back gardens.

The football training worked and we retained the Infantry Cup in a 6–5 goal thriller. I could not resist wording the after match signal to the battalion '1 King's Own Border 5 (five) – 3 R ANGLIAN 6 (six)'. When the signal was tapped out on the teleprinter in Creggan Camp and the figure '5' appeared, there were groans, only to be replaced by cheers seconds later. It really is important for soldiers to win.

Terrorist activity did not diminish during our last month in Londonderry. On 4 July two of our bases were attacked simultaneously. In the follow up there was a major gun battle, during which Pompadours fired some 200 rounds. It extended for a brief, violent period of street fighting, during which we killed two leading terrorists.

By the end of the month we had returned to Paderborn, having again handed over our patch to our 2nd Battalion. For the rest of the year, the 1972 blueprint was followed – block leave; followed by a rapid re-introduction to armoured soldiering; another divisional FTX and a major exercise in Denmark; before Christmas loomed again.

For Sue and me, the block leave was spent in Italy, in a newly acquired, albeit second hand tent, at the Lido di Jesolo, in company with Alan and Ros Behagg. Alan had been the Intelligence Officer in Londonderry and we both had young families. The campsite was extremely full and the only tent sites available were immediately outside the washing and toilet block. Never one to miss a quip, Alan joked that we had moved from the Creggan to the Bogside – and to think that I thought Paul McMillen's jokes were awful!

I was coming to the end of two years as adjutant and it was time for me to move on. Jonathan Hall-Tipping was determined that I should get a good posting and was instrumental in securing my next appointment, as the regimental representative and an instructor at the Royal Military Academy, Sandhurst. Once the appointment had been confirmed, I was sent to Cyprus, where Sandhurst's existing intake of officer cadets were undergoing their final exercise before commissioning. The idea was that I would get to know the other instructors in the company that I was to join and see how officer training had developed since I had completed it in 1963, however, the exercise that I observed looked remarkably familiar.

I returned to Paderborn, but left the battalion mid-way through the divisional FTX, having handed over to an officer from the 2nd Battalion, Bob Alpin. After that it was a matter of packing, handing over the quarter and getting Sue, Nick and myself to Camberley and settled before Christmas.

CHAPTER NINE

SANDHURST – A DECADE LATE

I HAD FAILED TO get into Sandhurst as an officer cadet, but after a delay of some ten years, I was finally there, as an instructor – better late than never.

We travelled by car from Paderborn to Camberley. As I drove our car off the ferry at Dover, Nick, on entering England for the first time, said his first, unmistakably clear words – 'Oh dear!' Hopefully, not a harbinger of things to come.

The first thing on the family agenda was to sort out a quarter. I went to Sandhurst's Families Office to be told that I had a choice of two. That appeared to be good news, but the look on the face of the estate warden told me otherwise. He informed me that one quarter was in dire need of redecoration. In recent months it had been pressed into service to meet an emergency and, consequently, it had missed its slot in the painters' schedule. The other quarter was better decorated, but had no central heating. Sue and I agreed to take a look.

We inspected the dowdy place first. It was awful. Its paintwork was chipped and scuffed and some of the upstairs rooms had paper peeling off walls and ceilings. Most of the soft furnishings were ancient and in dire need of replacement. Understandably, given its dismal decorative state, there was very little evidence of any 'march out' cleaning. 'No way' was our immediate response. We crossed the road to a house opposite. Admittedly, the paintwork was better, but it was like a fridge inside. Nick was only nine months old and he would need to be kept warm. Neither quarter was acceptable, but my protests to the Families Officer were in vain. There were no other quarters available. I could hire accommodation privately until the poorly decorated quarter was ready, but that was the only alternative. I asked for time to think about it. In the meantime, Sue, Nick and I could stay with Sue's Aunt Hilda, who lived in Penn Street, near Beaconsfield – geographically convenient and she had also looked after our cats while we were in Germany.

Eventually, we opted to be warm and set about pulling paper off walls and ceilings and deep cleaning the first quarter that we had been offered. Aunt

Hilda made us very welcome in her lovely home, while we were involved with scrubbing and polishing. It took several days of hard work to make the house habitable, but at least its heating was efficient. Once cleaning was completed, we spent a very happy Christmas with Aunt Hilda and her family. Festivities over, the three of us, now with our two cats, moved into 11 Matthews Road, Camberley, with painters promised early in the New Year and me ready to go to work.

Mons Officer Cadet School had closed in 1972 and its training responsibilities had been transferred to the Royal Military Academy, Sandhurst (RMAS). Up until that time all the colleges at RMAS had run identical two-year courses that combined military training with an academic syllabus, and officer cadets were commissioned on the completion of the courses. The closure of Mons was the catalyst to change the organisation at Sandhurst.

By 1973, RMAS had been organised into three colleges, each with different training objectives; New College catered for newly arrived officer cadets until their commissioning; Old College took over from New College to give newly commissioned officers some academic and follow-on military training; and Victory College ran courses for 'vicars and tarts' – an unjustified appellation for would-be Territorial Army officers and aspirants for commissioning into the smaller specialist corps, for instance, the medical and nursing corps, veterinary officers and chaplains.

I joined Normandy Company in New College and was to become 4 Platoon's commander/officer instructor. My company commander was Major Nick Lawson of the Lifeguards. His second-in-command was Lieutenant Commander Peter Libby, the Royal Navy's exchange officer at Sandhurst. The other two platoon officers were Jeremy Dumas and Mike Wilcox, both Royal Artillery officers. The three platoon SNCO instructors were all drawn from Guards regiments, because of their expertise in drill and the exemplary standard required at Sandhurst for its parades. My SNCO was Colour Sergeant Derek Rossi, Grenadier Guards and he and I got on like a house on fire from the outset. The final SNCO, who looked after the company's administration and was none other than Colour Sergeant Graham Taylor, the Pompadours – the first soldier I had met on arrival in Belfast in 1963. Our career paths were advancing quickly and running in parallel.

I was keen to see exactly how much of the Mons syllabus had been retained.

In fact, I cannot recall any significant change having been made at all. That the first five weeks of the twenty-week course were devoted to basic training, was entirely predictable, but the fact that the remainder of the course was near identical to that of the early 1960s was remarkable. The Mons syllabus and practices had been transferred to Sandhurst lock, stock and barrel. Even the exercise names were the same and *Exercise Marathon Chase* was still the first tactical field test faced by officer cadets. It continued to generate the same apprehension in them that it had done for me and my contemporaries a decade previously.

The new intake of officer cadets was divided into two companies, each with three platoons. I poured over the list of names of the officer cadets who I was about to receive, together with their reports from the Regular Commissions Board. Those reports were detailed and uncannily accurate with regard to how well the RCB assessors thought various individuals would react to training at Sandhurst. There was no doubt that the accuracy of those reports coloured our thinking, so much so that the practice of letting instructors see them before the courses had got properly underway was discontinued very quickly.

Even in 1974, officer cadets were drawn from a wide band of civil society. In my first platoon I had a hereditary lord, an Etonian, who became very friendly with a tractor driver's son – who had been to Welbeck College, the Army's boarding school for aspirants wanting to join one of the technical corps. There were also a couple of fast track, British Army NCOs recommended for officer training and a clutch of overseas cadets. I got to know every one of my officer cadets' backgrounds extremely quickly because they were required to write an essay about themselves during the first weekend they spent at the Academy. Most were amazingly candid and, coming from a very stable family, I was surprised at just how many had suffered some significant disruption in their upbringing.

Every week I held a progress review with Colour Sergeant Rossi. By the end of basic training we had a pretty good idea where the strengths and weaknesses lay in the platoon. As major milestones were reached in training, the reviews were held at company level and the warning system came into play for the least successful. Most mentoring came in the form of communal debriefs at the end of an activity or exercise, but I also employed informal, one-to-one sessions that could take place in my office, but they were just as likely to be conducted anywhere, as opportunities presented themselves.

The major part of my job was lecturing and observing all aspects of the cadets' training, but as tactical training progressed, the scheduling of command appointments required considerable thought. The trick was to judge just how much an individual could cope with, without the command task running beyond his abilities. Success would often accelerate progression; a moderate performance could become a platform to build on; but a marked failure invariably meant some major re-building of confidence would be required.

Just as it had been at Warminster, at RMAS an instructors' life was regulated by the training programme and there were few surprises. However, I was about to get one. Princess Anne had recently married a fellow equestrian and Olympic gold medal winner, Captain Mark Phillips. They were to begin their married life at Sandhurst and there was speculation in the New College officers' mess as to which college and company he would join. The balance of opinion favoured Old College. The pace of life there was much less frenetic than that in New College, where, after basic training had been completed, most weeks contained one or two night exercises, or a period of continuous training that lasted anything from two to five days. We could not see Mark Phillips being pitched into that, with royal engagements to contend with as well his military duties. In the event, it was Normandy Company that would receive him, moreover, he would shadow me for the remainder of the current course, before taking his first platoon of officer cadets through their training. Apparently, the thinking behind this decision was that the Army did not want to make Mark's Sandhurst posting appear a sinecure and being a New College instructor certainly avoided that pitfall; Nick Lawson was a Household Division officer, so that sat well with Princess Anne about to be in close proximity; and the course was infantry based and I was the company's infantry officer – all those factors linked neatly to put Mark and myself together.

I first met him on the parade ground in front of New College. Colour Sergeant Rossi was there with a handful of cadets from our platoon to provide a backdrop for the waiting gang of press photographers. Mark was getting used to this level of attention, but it was entirely new to the rest of us. The results were evident on the television news that evening, but the following day, my parents, as avid readers of the *Daily Telegraph*, were delighted with a big picture of Mark and me on one of its centre pages.

By the following day the company had left Sandhurst for a major exercise on Dartmoor. Colour Sergeant Rossi, Mark and me had dropped off small parties of cadets around the periphery of the moor. Their first task was an approach march of perhaps 12 miles to a rendezvous point, carrying about 40 pounds of weaponry and equipment, across difficult terrain. Each party had a different route, but their routes converged at a mid-point, so that their progress could be monitored. That was where Colour Sergeant Rossi, Mark and I headed. We got there well in front of the labouring officer cadets and Colour Sergeant Rossi suggested that we have a brew. Mark was carrying the teabags. Colour Sergeant Rossi moved behind him to get them out of his webbing's back pouches.

'For fuck's sake, sir, stand still,' he said. Mark burst out laughing.

His response was, 'That's the rudest thing anyone has said to me since I got married!'

I had wondered how I would get on with this high profile cavalry officer. The few apprehensions that I had entertained quickly disappeared. He was a down to earth bloke and a good soldier, despite the 'Foggy' nickname that was often attached to him by the press.

I did not find my instructor role particularly challenging. It took some time to prepare the lessons or lectures I was required to deliver and monitoring the progress of my officer cadets was vital as their military abilities and personal attributes developed – or in a few cases, failed to do so. But what I did find demanding was the preparation for the staff and promotion examination that was to be held later in the year.

For the majority of officers, the staff and promotion examination is one of the biggest hurdles they have to confront in their careers. The examination covered three subjects: military administration, war studies and international affairs. A mark of fifty per cent or above, in each subject, secured a staff pass and brought consideration for a place at Staff College. A lower pass rate, again in all three subjects (forty-five per cent I think), was sufficient to gain eligibility for promotion to major.

Only a relatively small percentage of officers are selected to attend Staff College and those who make it know that they are in the upper quartile of their contemporaries and are destined for a successful career. Eventual promotion to lieutenant colonel is virtually assured, with the likelihood of further advancement for many. Conversely, however successful you had been in your career up to that point, it counted for very little if you fluffed the staff examination. You would be back swimming with the mainstream, so the pressure was on and, as my examination record was woeful, I was nervous that it might continue in that vein.

The preparation required quite a lot of reading and attendance on week-long, residential courses that covered war studies and international affairs. I attended four of them, held at Southampton University, with lecturers drawn from the RAEC, supplemented by visiting civilian speakers. They were well run and very informative. During the courses we wrote essays against the clock, in simulated examination conditions. I was heartened by the good results I achieved. It was down to personal effort and detailed swotting to prepare for military administration, which included a working knowledge of military law.

The two days of the examination arrived with three, three-hour papers to negotiate. Military administration came first, for which we were allowed to bring our carefully amended Manuals of Military Law – a reference work widely used by the civilian legal profession as case law is brilliantly recorded and simply explained. I think war studies came next and the paper provided no particular alarms. I was reasonably satisfied with my first day's work. Certainly, I was confident of a good pass in military administration as I taught a fair part of the syllabus at Sandhurst. The following morning I settled down to read the questions posed by the international affairs paper. There were four sections with one question to answer from each. Three posed no problem, but the fourth section was a stinker. I left it to last and wrote, or rather invented, something on Chinese foreign policy. It was all in the lap of the gods now.

There was a wait of several months before the results were published. I was the only captain instructor at RMAS not to secure a staff pass, but my results were remarkable. I achieved over 80% for international affairs (obviously the Chinese had somehow taken heed of my suggestions) and got more than 75% for war studies. My mark for military administration was 48.5%! I could not understand it. Military administration was factual stuff and I was confident of my knowledge. However, 50% in all three subjects was the requirement and I had not managed that, so a promotion pass was my meagre reward. I would have to sit the examination again next year. It was a real downer.

As the commissioning course progressed, my role as the regimental representative became increasingly important. Officer cadets still had to list three choices of corps or regiments they wished to join. Once they had done that I interviewed all who had The Royal Anglian Regiment amongst their choices, with a view to gauging their suitability and motivation to be an infantry officer. I had no authority to stop anyone applying to join my regiment, but I could exert some influence if I felt that they were making a mistake – for instance, quite often officer cadets would apply to join their father's regiment or corps, when they were very different people, with different interests and skill sets that would be better suited in another regiment, or service, or (whisper it) even in civilian life.

With initial interviews completed, I would draw up a list of candidates for the Colonel of the Regiment, or his senior representative, to interview more formally. The second interviews would produce acceptances and that meant that the officer cadets selected would almost certainly join the regiment, once they were commissioned. Other regiments, or corps, were not allowed to poach once acceptances had been confirmed, but an officer cadet was at liberty to change his mind if he decided that in the light of further experience, he would be better off elsewhere. My job was to keep those who the regiment had accepted, well briefed and involved in what our various battalions were doing and what their immediate futures, on leaving RMAS, might bring. It was vitally important that I secured sufficient officer cadets to keep the regiment's officer manning healthy.

1974 was a year of global recession. In the UK a three-day working week was introduced; key sector strikes were commonplace; a general election resulted in a hung parliament; and following the Arab/Israeli war of the previous year, OPEC imposed an oil embargo that caused petrol shortages, made worse by panic buying. The petrol shortage was the problem most keenly felt by the Army and RMAS was no exception. Petrol for the Academy's transport fleet was severely rationed. Much of it was reserved to support major exercises, many of which were held on relatively distant training areas e.g. the Brecon Beacons, or Stamford in Lincolnshire. Local exercises were not allocated vehicle transport, but Sandhurst had a solution.

Before the Mons course was transferred to RMAS, officer cadets were provided with heavy duty bicycles to move them quickly around the very considerable estate that houses the Academy. I have no idea why the practice was discontinued, but the bicycles had been retained. They were desperately needed now to facilitate local exercises and keep the syllabus going. However, putting as many as 200 cyclists on the road at the same time, would create a serious traffic hazard, so some organisation was needed. Step forward Sergeant Major 'Wullie' Fullerton, Scots Guards. He was older than many of his SNCO contemporaries and he remembered bicycle drill. Watching officer cadets drill holding their heavyweight velocipedes was a scene worthy of Monty Python, but it got them into pairs and able to stop and start together. A major road hazard had been mitigated, but 'Wullie' was not finished.

He was the instructor for No 1 Platoon and that meant his platoon was at the head of the bicycling column. 'Wullie' realised that not all of his overseas cadets could ride a bike, so he put them at the back of his platoon and waited for the first bicycling expedition to get underway. The column formed in its pairs and was told to 'Walk march' (a cavalry order) then 'Prepare to mount – mount'. With that the cadets swung their right legs over their saddles and started peddling. The bulk of No 1 Platoon cycled off happily, but there was a lot of wobbling in its rear files before the inevitable pile up occurred. No 2 Platoon had nowhere to go and added bodies and bikes to the carnage in front of them. Subsequently, there were mini-pile ups throughout the column, which made for an incredibly funny, farcical, never-to-be-forgotten scene.

Sport was taken very seriously at Sandhurst. Everyone participated and the various representative teams played their sports at an extremely high, amateur level. I became the Academy's football officer and was helped in this task by a professional coach, Ron Bell, a former Chelsea player. Occasionally I turned out for the Academy side, but normally I deferred to goalkeepers drawn from officer cadets and newly commissioned officers attending Old College because playing for the Academy was good for their self-esteem. Most of my football was played with the Army Crusaders – a team comprised entirely of officers and drawn from across the Army. It used Sandhurst for its home fixtures and I enjoyed two seasons of good football with them. Athletics was already a thing of the past and my running was purely to maintain fitness.

The course continued with little to remark on except for one anti-terrorist exercise that required officer cadets to mount road blocks within Sandhurst's grounds. Instructors' wives were persuaded to provide passing traffic and Sue was included, complete with baby Nick in his car seat. It gave the cadets a dimension to deal with that they had not anticipated. What they certainly did not reckon with was Princess Anne, in full flow, when asked to open the boot of her car! A considerable amount of moral fortitude was required of them to cope.

At the end of the course, the Sovereign's Parade saw the commissioning of the successful officer cadets. The backdrop for the parade was the impressive façade of Old College. The cadets were dressed in 'Blues'; a major band was always engaged to play and augment the RMAS band; and the standard of drill was immaculate. Sovereign's parades were always wonderful occasions. Parents came from far and wide to watch their sons and daughters slow march and follow the Academy adjutant, on his white charger, up the steps of Old College to the sound of 'Auld Lang Syne'. Moving stuff and tears were guaranteed. The parade was followed by lunch, before the boys were off to London for the commissioning ball, then held at one of the major hotels.

Normandy Company's officers and their wives were particularly lucky to be invited to Princess Anne's private apartment in Buckingham Palace before going on to the ball. Sue and I arrived at the Palace's right hand gate in our highly polished Austin Maxi. The policeman on duty indicated the archway that I was to drive the car through and opened the gate. The two Guards sentries came to the salute as we passed them and we parked in an inner quadrangle. A footman appeared and showed us to the apartment. We were 'living the dream'. Drinks were pleasant, informal and entirely comfortable. Princess Anne's private staff knew her well and there was a particularly warm relationship between the Princess and an elderly lady, with a broad cockney accent, who had probably been with her all her life.

The party moved on to The Dorchester and the rowdy noise made by relieved, soon-to-be officers in their mess kits, with tape over the single star on their shoulders. Their commissions would become active at midnight and the tape could come off then. In the meantime, dinner was served. At one point a fly attempted hari-kari in Princess Anne's wine glass. Mark called a waiter over, who apologised profusely, disappeared, only to return seconds later with a napkin and silver tea spoon. He lifted the glass, used the spoon to remove the fly to the napkin and placed the glass back in front of HRH! He smiled a wide smile at everyone at the table and made an elegant exit. Not the service she got at home.

The final highlight of the evening came when Sue and I took a taxi back to the Palace to retrieve our car. 'Buckingham Palace,' I said to the driver. 'Right oh,' was the reply and when we got to the Queen Victoria Memorial in front of the Palace, he wanted to know where to go next. 'End gate on the right,' I told him. I had been saving that up ever since we had got into the taxi and he appeared to be the happiest man in London when we were inside the Palace's precincts – 'Wait till I tell the missus!'

A few weeks later my second course formed up. I had another Grenadier Guardsman, Colour Sergeant Hobbs, as my platoon's SNCO. He was very capable and a sound judge of cadets' characters. The course would run smoothly, except that the final exercise posed a big problem. The exercise was normally held in Cyprus, but Turkey had invaded the island in the previous July to establish a major enclave in the North. Displaced Cypriots, both Greeks and Turks, were still crossing a newly established United Nations buffer zone to whatever was safe territory for them. The country was in such turmoil that Sandhurst's final exercise could not be held there. A new exercise had to be written for another location – Malta. I was warned that I would be required to help write it, but that was several months away.

Mark and I had developed a good, friendly relationship. I admired how much effort he put into his instructing role, while he maintained his position as a leading British three-day event rider and partner to Princess Anne at high profile events. I did the little I could to help. Very occasionally I would go his quarter, Oak Grove House, a large property inside Sandhurst's perimeter that was suitable, particularly from a security point of view, to be home to a member of the royal family. Mark and I would go over exercises and various lectures and lessons. In the process I met Princess Anne and that is how she learned of Marks and Spencer. Both cats were great hunters and Princess Anne and Mark had a plague of mice in their stables. Could Marks and Sparks help? Certainly they could. They became 'by Royal appointment' moggies and made several successful forays to the stables to decimate mice numbers.

The Dartmoor exercise came around again. However, Princess Anne was due to fly to Canada a few hours after the exercise finished and Mark was required to accompany her. He was lifted off Dartmoor by helicopter and flown to Camberley before being whisked across the Atlantic, only to teach at RMAS again three days later. His was a punishing schedule. He had taken his car, a very powerful Rover V8, to Dartmoor and asked if I would drive it back to Camberley. Of course I agreed, but when that became known, other Normandy Company officers wanted to 'bag' the spare seats – so much more comfortable than a Land Rover.

We set off on narrow, West Country roads and I was stuck behind slow moving traffic for miles. After the rigours of the exercise all my passengers fell asleep quickly. Finally, we reached the Honiton by-pass with its dual carriageway. I pressed the Rover's accelerator and the car surged forward. I was at the front of the queue in seconds and enjoying myself when a figure, several hundreds of yards in front, stepped into the road to waive me down. A traffic cop!

'You're in a hurry, sir. Did you realise you were doing nearly 100 miles an hour?'

Before I had a chance to reply, the police radio in Mark's car began to squawk. 'Why have you got a police radio?' said the policeman. I explained that the car belonged to Mark Phillips and Princess Anne.

'Really?' he said. 'Can you prove it? Have you got insurance?'

I told him that I could only prove my identity and those of my fellow passengers and that we were all Sandhurst instructors. Our ID cards seemed to satisfy him, but he gave me a 'ticket' and said that I would have to produce proof of insurance at my local police station. I was never able to produce the required proof. Crown indemnity was still in force at that time and the Royals and their household were able to drive without insurance – apparently I qualified as a temporary member. Sadly, the indemnity did not save me from a hefty fine.

Sue was pregnant with our second child. Tim was born in March at the Princess Louise Margaret Military Hospital in Aldershot. The birth was traumatic in several ways. Sue was taken into hospital the day before she was due to give birth. She was induced early the next morning, timed so that she would be ready for the gynaecologist when he arrived for work. However, the fact that she was placenta praevia had not been detected. Now the baby had been induced and was on the way, there was an emergency to address, because if all did not go well, both mother and baby might bleed to death. I was telephoned at home, at breakfast time, by a young nurse and was asked not to come to the hospital, but to wait until the emergency procedure had been completed. I took Nick to Jeremy and Liz Dumas (I worked with Jeremy and they were near neighbours). Liz made Nick and me a cooked breakfast. Jeremy gave me a stiff whisky. I appreciated both. After breakfast Nick and I made a very fine snowman and that kept us occupied until the hospital rang again. Liz looked after Nick and I hot footed it to Aldershot.

Tim had been delivered by caesarean section, so Sue was still out for the count. I was taken to see my second son. Instead of the blond bundle I had anticipated, when he was handed to me I was holding a tiny baby with inky black hair and a dark olive skin. I was shocked. 'Are you sure this baby is ours?' I asked. I was shown a little plastic tag on one of his legs with 'Groves' written on it. I was assured that there could be no mistake because the tag was put on only seconds after the birth. I was still not convinced, particularly as a Ghurka battalion was stationed in nearby Church Crookham and this baby looked to have much more in common with those Nepalese warriors than he did with me. The gynaecologist arrived to tell me that Tim was badly jaundiced and that the skin colour would fade as he recovered from the condition. I simply had to accept his word. However, nothing could alter the fact that I had two wonderful sons. Sometime later, when they had fallen asleep together, their relaxed faces looked uncannily alike, except one was blond and the other was dark – and still is, given a few grey flecks, more than forty years later.

Within six weeks of Tim's arrival, on 6 May 1975, my father died. It is a deep regret that he and I did not know each other better. When I was a child he was a somewhat remote figure. I think he may have had an over-developed sense of responsibility because it seemed that his business, and once that had been sold, his work, always took priority – maybe it was his way of ensuring proper provision for his family. Nevertheless, I found it hard to forgive him for opting not to attend my commissioning parade. That hurt. At the time it was the most important day of my life and an occasion when I wanted my parents there to celebrate with me. I never confronted him with his absence and our relationship continued much as it had before – that is until the last time I saw him. I had spent a weekend in Sudbury at my parents' home. My father was confined to bed and I went to his room to say a normal 'Goodbye'. As I entered, his face lit up. He lent forward from his pillows and seized my hand in both of his. He thanked me for coming, and said how proud he was of my young family and that my career was progressing so well. On leaving, I turned to give him a wave and he was smiling a warm, happy, loving smile. It is a wonderful last image of him and one I treasure.

At Sandhurst the commissioning course was nearing its final stages and I was flown to Malta, with the College's chief instructor, to write the final exercise. We had four days to complete the job. Malta had a military garrison and so training areas were available, but they were narrow strips of land divided from one another by towns, villages and farmland. It was difficult to write an exercise with any flow. We surmounted the problem by including the small island of Gozo in our planning and so increased the real estate. Gozo had sufficient space for a defence exercise, while the separated training areas on Malta could be used to practise more mobile operations. Job done.

In less than two weeks the course was flown to Malta and deployed to the training areas immediately. Normandy Company was the first to use Gozo and almost as soon as we arrived there we were visited by an old friend, Lieutenant Commander Peter Libby. He had ended his tour at RMAS and had re-joined the Navy. He was now the first lieutenant (second-in-command) of a Type 42 frigate, HMS Charybdis, which was alongside in Malta's Grand Harbour. He invited the company's officers and SNCOs to visit the ship, when it was to conduct a brief sea trial in a couple of days' time. We gladly accepted and wondered what hospitality we could offer in return. We came up with the idea of hiring some donkeys and playing polo, with bass brooms as mallets and a football to make hitting easier. We issued a challenge to Charybdis' ward room.

We got a second invitation from the Senior Service. The Admiral commanding the Navy in Malta invited the company's officers to dinner at his official residence. The Navy live well. The Admiral's house was perched high above Valletta and overlooked the Grand Harbour. The view was magnificent and the house was imposing. We had an excellent dinner. When we had finished the meal and were relaxing over port and brandy, the Admiral turned to Nick Lawson and said 'I think it's a bit offside that you have challenged Charybdis to polo. You are a cavalryman, Mark is an Olympic horseman and I bet the rest of your officers have ridden since birth. The press have got wind of it and I don't want the Navy embarrassed.'

Nick explained that something had got lost in translation and our intention was to play donkey polo for the fun of it. The Admiral burst out laughing and asked for his Yeoman of Signals to join us in the dining room. A SNCO appeared and the Admiral dictated a short signal to the captain of the *Charybdis*, which lay in the harbour below us. The Yeoman of Signals walked to the end of the dining room and opened doors onto a veranda. On the veranda was a large *Aldis* lamp. The Yeoman turned it on, pointed it at *Charybdis* and worked a handle on its side. The lamp flashed *dit, dit, dit* – *dah, dah, dah* – *dit, dah, dit etc*. We joined him to see what would be the response. Within a minute or so *Charybdis'* light was flashing a message that read that they had been had, but wait until they got us on board!

In the event, the donkey polo was called off. Having the press present would have spoiled things, but the visit to *Charybdis* – affectionately known to her crew as '*the Cherry B*' – did go ahead. She was berthed immediately behind the *Ark Royal*, an enormous aircraft carrier, and it looked as if negotiating a way past her would be difficult. Peter Libby was in charge of getting his ship out of harbour. He had the engines started and ordered the forward lines to be let go. *Charybdis'* bow swung out and as soon as it cleared *Ark Royal's* stern, he let go the rear lines. If ships can leap, then the *Cherry B* leapt forward, clearing the harbour's entrance in no time at all and we headed, full speed, for Gozo. The sea was choppy and we began to suffer as the Navy had promised.

Like the rest of Malta, Gozo is very rocky with vertiginous cliffs marking much of its coastline. The frigate got closer and closer to the cliffs. At one point I could nearly stretch a hand over the rail to touch the rock and we were still making way. I asked the 'Buffer' – the senior seaman – 'What's he doing?' Even by military standards, the Buffer's reply was fruity!

After three or four hours at sea we returned to our company. The entire exercise ran its course as the chief instructor and I had planned it. Then it was the flight home in time to make final preparations for the commissioning parade.

This time the Sovereign's Parade was to be exactly that – the Queen was to take the salute and Sandhurst went into overdrive. RMAS was always an impressive place, but it was spruced up to the 'nines' because not only was the Queen coming, but the military 'great and good' of past decades were to be present too – literally dozens of generals and a couple of field marshals. Nothing but the best would be good enough. Captain instructors were allocated a general to host throughout the day. Mine was Major General Matthews, Colonel of the Regiment of the South Wales Borderers. Mark was made a temporary aide de camp (ADC) to the Queen.

On the day of the parade Mark and I were at work early to inspect our platoons and to be ready for our respective hosting duties. Mark came into my office clutching a piece of paper.

'Look at this,' he said, 'not one captain or SNCO is to be presented to the Queen.'

That really was a bit shoddy as those were the two ranks that shouldered the day-to-day burden of running the place. 'You're the ADC. You fix it,' was the best that I could offer.

General Matthews was a great character and in his eighties. He had been commandant at RMAS at the end of World War II and had wonderful stories to tell e.g. Old College is grey because he, personally, organised a night-time raiding party on a Naval paint store! He told Sue and I that he had officers' quarters designed at the end of the war, but the plans were refused and much cheaper houses were built.

'To add insult to injury,' he said, 'the buggers named a road after me. Where do you live?'

'11 Matthews Road,' I replied.

'Sorry' – and said as if he meant it.

When the parade was over, all the VIPs and hosts gathered in the hugely decorative Indian Army room, just inside the entrance to Old College. Champagne was served, but it was a tremendous crush. I was tapped on a shoulder and managed a half turn to look up to see Mark standing immediately behind me.

'Ah, Phillips, you old sod – how's your day been?'

I had not noticed a small woman standing to one side of him. She looked at me and smiled a knowing smile. 'Beam me up Scotty!' I was covered in embarrassment, but The Queen quickly put me at my ease. Only rarely does she experience anything that is 'off script' and she was genuinely amused. It was another, and by far the greatest, of my royal gaffes.

The year had progressed and the staff and promotion examination loomed again. I went through all the preparatory courses, mustering as much enthusiasm as I could, but it was a hard slog. I did spend additional time on military administration and felt reasonably ready by the time the examination came around. In the event, there were no great upsets, or great causes for hope. I simply did not know if I had done enough to pass at staff level. Another wait ensued.

It had not been a great few years for my parents and brother. My father had long suffered a debilitating affliction and had died of a stroke in 1975. My mother took her husband's death stoically and was fine. Prior to that, my brother had married in 1973 without informing his parents, or me. Rightly or wrongly, I regarded his action as a betrayal of love and trust. I had not even met his wife and felt badly let down. However, Sue had maintained touch with Rick and sought to rebuild relationships by organising a holiday together in Teignmouth, with our respective young children. We rented a house, but it was not a great success. Rick and Kate (his wife) had much more liberal ideas of raising children than we did and although arguments were avoided, the friction was palpable. Relationships were to remain cool.

The situation was not helped when I got news of my next posting. I had hoped for a staff job on a training team in Uganda and knew I was in the running for it. I had to ring Regimental Headquarters for confirmation while I was on holiday. 'Great career news,' said the voice at the end of the 'phone. Alarm bells rang – Uganda would be an adventure, but great career move it most certainly was not. The voice went on, 'You're going to be GSO3 Ops at Headquarters 8 Brigade in Londonderry.' Bloody hell! How do I break that to Sue? To top it all it would be another Christmas/New Year move. To translate the military jargon – GSO3 Ops means General Staff Officer Grade 3, working in the Operations department. I needed no more information to know that a very difficult posting lay in front of us.

By now Sue and I had decided that our family was complete and like other young couples at Sandhurst, our thoughts turned to getting on the housing ladder. Our immediate neighbours, Tim and Liz Robertson, had bought an end of terrace house in Frimley, but we thought that Surrey house prices were more than we could afford, so looked further afield. We knew that we wanted a house to rent and settled on Colchester – housing in Essex was considerably cheaper than in Surrey. Colchester is a university and military town, with a renowned school that taught English to foreign students. The town is also in commuting distance of London. All of those were significant pluses as far as renting was concerned. We contacted several estate agents and made a shortlist of six properties that we wanted to view over a weekend. We would stay with my mother, in Sudbury, while we went house hunting.

None of the six selected properties matched our expectations, but we found another house that looked ideal. It was detached, with three bedrooms and an integral garage, all for the princely sum of £12,700. The problem was that we had not made an appointment to view it. The owners were out and we had to leave a note asking if we could call on the following day. Next day we found another note informing us that the owners would be pleased to see us, but they would be out until about 6 o'clock in the evening. We duly arrived at what we thought was the appointed time, but no one was at home. We waited until 8 o'clock and then gave up. As we were driving away, a car drew into the driveway. We received profuse apologies and were immediately shown over an immaculate house. It was a perfect rental property. Sue and I had a brief conversation and offered the asking price, but only if all the carpets and curtains were included.

'Done,' said the owner. 'Let's have a drink and then see what else we can throw in.'

The owners were extremely generous and carpets, curtains, white goods and some furniture was included in the selling price. Better still, the house was ours only six weeks after our first seeing it.

Shortly before leaving Sandhurst, I got a telephone call from Sue while I was at work in my office. Nick, aged two, was missing. He had a sit-on orange tractor and yellow trailer and could really make the thing move. He had been playing in the drive at our quarter and suddenly he was no longer there. Sue had searched the quarters' area, but there was no sign of him. My immediate worry was that the ranges were active and he could have gone into the danger area, which adjoined the quarters. I rang to stop the firing and the officer cadets were organised into search parties. I went home to talk to Sue and to join in the search.

Most of the Sandhurst officers' quarters are located about a mile from the Academy and the route between them runs through some pleasant woodland. It was a favourite walk for many RMAS families. One weekend, a week or so before this incident, I had taken Nick to my office because I had forgotten to bring some papers home. While we were there I produced some pencils and paper for him to draw on, then located what I needed and did a few minutes reading, ten minutes at the most. Apparently that had been enough for Nick. Now bored with trundling up and down our driveway, he thought that he would come to see me. Unerringly, he made his way along the road, through the woods, across Victory College's parade ground and up a rise onto the parade ground in front of New College. Rhine Company were being drilled by their company sergeant major. Undeterred by all the shouting and foot stamping, Nick, on his tractor and trailer, scooted between the sergeant major and his company and disappeared into New College. Little legs pumping, he made his way along the central corridor, found my office, got out the pencils and paper and settled down to draw.

There was no sign of Nick in the danger area, so I returned to my office to call the police. I opened the door – 'Hello Daddy'. Relief flooded over me. I rang Sue with the good news and the search parties were stood down. Nick has always possessed the most extraordinary sense of direction and memory, but we were not aware of those talents until that remarkable, albeit alarming feat for a toddler.

In the week before my tour at Sandhurst was complete, I hired a large van and with the considerable help of Tim Robertson, moved some of our furniture to our new house in Colchester – we had, piece by piece, purchased and/or renovated items of furniture during our postings in Warminster and Camberley. We were in our own home by Christmas and Sue would spend several weeks into the New Year in Colchester because no quarter was immediately available for us in Londonderry.

Our final night at RMAS was spent at Oak Grove House, where Princess Anne and Mark held a dinner party for us. The company and conversation was very 'horsey', but it was a kind and memorable ending to our time in Camberley.

CHAPTER TEN
HARD TIMES IN LONDONDERRY

THE YEAR 1976 was only a few days old when I flew to Northern Ireland, leaving Sue and the boys behind in Colchester. On arrival at Aldergrove Airport I was met by Ian Jones, the officer I was to replace. He drove me to Londonderry – the long way round. The direct route was across the Sperrin Mountains, but that ran through the small towns of Maghera and Dungiven. Both were strongly republican and their populations and those in the countryside surrounding them, were broadly sympathetic to the IRA. Breaking down, or being stopped, anywhere in the Sperrins could, in the wrong circumstances, be fatal. We drove north to Coleraine and then west to Londonderry; it was an early introduction to the many inconveniences that would combine to make life difficult over the next two years.

I settled into a bedroom in the officers' mess in Ebrington Barracks, situated on the east side of the River Foyle. The barracks was home to a resident battalion, 1st Battalion, The King's Regiment, but Headquarters 8th Infantry Brigade was also located there. 'Resident' meant that the battalion would serve for two years in the Province and married soldiers would have their families with them. Resident battalions were not allocated large, permanent areas of responsibility. Instead, whenever the security situation demanded it, they would send companies to reinforce units that were tasked with controlling difficult areas.

The following day I reported for duty. The headquarters compound was housed in a separate secure area within the already secure barracks. Its two-storey, flat roofed building had a tall metal fence surrounding it, topped with razor wire coils. The entrance was guarded and admission was closely controlled. The clerks' and communications domains were on the ground floor. The first floor contained the brigade commander's and staff officers' offices and, most importantly, the operations room. My cramped office adjoined the operations room and was shared with another grade 3 operations officer. His principal job was community relations and liaison with various groups across the political and religious divides. Another grade 3 operations officer ran the operations room and he had the most interesting job. Disappointingly, my own was a little bit hum drum.

Much of my time was taken with arranging visits for VIPs – mainly politicians on fact finding tours and senior military officers. They formed an endless stream. I was also responsible for arranging the perimeter guards at Magilligan Prison, located on a promontory on the north Antrim coast. The prison housed many convicted terrorists, just as the Maze Prison did, but Magilligan was smaller and much less well known. The RUC had assumed primacy for the security situation (i.e. it led on deciding security responses, but still relied heavily on the Army to carry out virtually all of the operations). I became the secretary to the Divisional Police/Army Liaison Committee, working directly to the RUC's divisional commander. Finally, I was the dogs' officer! That was a fun job.

The brigade had a specialist dog section, of about twenty animals. Day-to-day the section was run by a warrant officer, but a staff officer was needed to represent its interests. To do that effectively, I had to learn about the foibles of the various types of working dog.

Guard dogs were generally huge and nasty – mainly Alsatians and Rottweilers. They guarded various installations and seemed to hate everything and everybody, other than their handlers. I maintained only a passing acquaintanceship with them.

Search dogs varied, with spaniels probably the dominant breed. Like sheepdogs, they loved to work and got very excited when they were put into their working harness. They were employed mainly on searching buildings and vehicles, looking for weapons, explosives, drugs and other evidence such as the clothing of a suspect implicated in a terrorist act.

Sniffer dogs were the divas. Like search dogs, sniffer dogs were employed to find things, but their speciality was explosives. They sniffed out car bombs and other IEDs. They liked to work, but there must have been an old 'Red Robbo' trades unionist type amongst them, because they would sometimes tire inexplicably quickly and this was dangerous, because if a dog did not indicate accurately, the troops with it might well walk into an explosion. Once back from a task, however short it might have been, sniffers would often resent being called out again if their rest had been truncated. Again, it was the lack of accurate indications that posed the danger and sometimes their handlers had difficulty in explaining this to officers who were commanding clearance operations. That was where I could help. The handlers (private soldiers, lance corporals and corporals) would ask a commander to get in touch with me to confirm how dangerous it would be to work a tired dog. I always understood a commander's frustration that an operation had to be abandoned, or curtailed, because a furry, four-legged mutt was having an off day.

I lived in the officers' mess for over two months, during which time I became familiar with my new role and the brigade's area of responsibility. It was huge, encompassing the whole of County Londonderry, most of County Antrim and a small piece of north Tyrone. Terrorist activity was not great, but it was unrelenting. Almost every day there was, at least, one bombing, or shooting incident somewhere on the patch.

Londonderry was, by far, the most active area and the most difficult to control. I made visits to all the units the brigade commanded to make myself known and to meet the key personalities. To bring myself up to date with conditions on the ground, I joined patrols in the Creggan, Bogside and the Shantallow – all republican areas in the City. And to gain insight into the rural areas, I took the operations branch Mini, a green car with civilian number plates, to drive around in bandit country, with a guide from whichever unit was responsible for the area. Both of us would be armed with 9mm Browning automatic pistols, but the car was not equipped with a radio. The safeguard system was to telephone ahead to the next location to be visited, saying what time we expected to arrive and what route we were taking. If we failed to turn up, they would come looking for us. It was a far from foolproof system, but the best that could be done in the circumstances.

In the second half of February, an officer's quarter became vacant at 8 Ebrington Park, only a few hundred yards from the barracks. Despite its closeness, journeys to work were always made by car. Walking was regarded as too dangerous. Ebrington Park had been made into a cul-de-sac with large concrete blocks preventing access at one end of the road that extended some 250 metres. At the other end, which led directly onto the A2 trunk road, was a sandbagged sentry post. The sentry there controlled a barrier and access into and out of the quarters' area. There were in the order of twenty quarters on the 'patch'.

I arranged for my furniture to be moved from store to Londonderry so that I could set up the house before Sue and the boys arrived. Eventually, a Castle's

Removals van and crew made their appearance and I joined them outside No 8. As the removal men began to unload our relatively few possessions, one said to me, 'I've never been to the Creggan and Bogside before.' I asked him what he meant and he explained that they had several part loads on board and that they had just delivered to two addresses, one each in the Creggan estate and the Bogside. STOP, STOP, STOP! Inside the van were MFO boxes that Sue and I had utilised to pack some of our smaller possessions. Clearly stamped on them and visible from outside the van, was 'CAPTAIN GROVES, 3 R ANGLIAN'. I asked if anyone had had access to the van while they had been in the Creggan and Bogside. The crew could not be certain as they had left the van for short periods to deliver various items to their customers' houses. I had to call for a search dog to clear my possessions before any further unloading could take place. Very little was straightforward in Londonderry.

In early March I flew back to England, to meet Sue, the boys and the cats at her parents' home. Our house in Colchester had been let. With the Maxi loaded to the gunnels, we set off for Liverpool and an overnight ferry crossing to Belfast. Nick was excited; Tim was too young to know what was going on; the cats were not best pleased to be confined in travelling baskets; and Sue was worried, but doing her best not to show it.

On arrival at Liverpool docks, my car was selected for an in depth search, including it being raised on a car lift so the searchers could poke about underneath. I showed my military ID card to the man in charge. It made not a jot of difference, despite him acknowledging that I was a member of the security forces. The crossing was uneventful, albeit not very comfortable in a grotty, cramped, overnight cabin. Once we were reunited with our car, we began the journey to Londonderry taking the long way around. Travelling through some of the Protestant areas of Belfast, Sue quickly noticed their loyalist murals and red, white and blue curb stones. Almost immediately we were behind an Army lorry with armed soldiers looking out over the tailgate. They were obviously very alert, with their rifles held ready. Sue found the whole environment extremely disconcerting and it was good to leave Belfast and to travel into the beautiful Antrim countryside, where the inter-sectarian threat was much less evident.

Sue was welcomed to Londonderry, particularly by our excellent next door neighbours, Tony and Gill Northey on one side and Nick and Denise

Williams on the other. They, and the other inhabitants of Ebrington Park, made life bearable for the next eighteen months. All of us, but particularly the wives, placed almost total reliance on one another for day-to-day general support and entertainment.

Almost as soon as Sue had finished her welcome cup of tea, I had to brief her on security. I showed her how to use a mirror, on a telescopic handle, to look under the car – a 'must do' every time she intended to use it and to describe, as best I could, what might be a suspicious object. Next came the dustbins – old style, round, metal bins. We scribed a pencilled line carefully so that it just marked the edge of the lid and the body of the bin. If the line was not absolutely aligned, then the lid had been moved and it would have to be cleared by an Ammunition Technical Officer (ATO) – the chaps who defused bombs – or a search dog, before it could be safely used again.

I described restrictions to her movement. Locally, she had free run of the east side of the city, except for the republican Shantallow estate. She could also cross the Craigavon Bridge to the main part of the city, provided that she stayed in the principal shopping area, most of which was contained within the city's walls. Beyond the walls lay the Bogside, which was a serious no-go area for our families. More widely, she could not venture north – bizarrely, southern Ireland lay in that direction. West also led to the Republic and to the south lay the Sperrins and bandit country, so both were out of bounds. She could go east towards Limavady and Coleraine, but there was very little to do in these small, sleepy towns. It took her no time at all to realise that her horizons were virtually restricted to coffee and tea with neighbours in Ebrington Park; to the one small supermarket in the east side of the city; to St Columb's Park, which, thankfully for a mother with two tiny children, was only a five minute walk away; to very occasional shopping outings to the city on the west bank (without the children) where a huge variety of bomb-damaged goods were good buys; and weekend trips, as a family, to the east, either to the swimming pool at Shackleton Barracks, Ballykelly, or to locations on the north Antrim coast – safe territory where most places did not witness a terrorist act throughout the 'Troubles'.

A few weeks after my family had joined me, the staff and promotion examination results were published. My clerk knew how important they were to me and came to my office to say that they were being transmitted on a teleprinter in the communications centre. I went quickly to the communications room and stood by the printer as it stuttered out the text line by line. The results came in alphabetical order and it seemed an eternity until they got to 'G'.

What an enormous relief. I had my staff pass and, in doing so, I had jumped a significant career hurdle. The results also contained the information that I had been selected to attend a staff course, first at Shrivenham and then at Camberley, commencing in October the following year. Shortly after I learned that I was to be promoted to major at the earliest opportunity. Great news all round. Sue was pleased for me and delighted that my forthcoming majority would mean an increase in pay.

When I got my detailed staff examination results, it turned out that International Affairs was a modest 57%, War Studies 52%, and Military Administration 50%. Military Administration was right on the pass mark, which almost certainly meant that my paper had been re-marked. I had cut things really fine and to this day, I still fail to understand how I made such a mess of the Administration papers in both years. But that was not the end of the matter. A little later I received a letter that brought even better news. I was no longer going to Camberley, but instead would attend staff training in ... Australia! Selection to represent the British Army at a foreign staff college was a huge feather in my cap and Australia held promise of a tremendous adventure. That news came on a chill, wind-swept, rainy day in Northern Ireland, but the sun was shining on me and my family. Sue and I could hardly believe our luck.

At home, Sue was adjusting, as best she could, to her new life. I was not there to help much. The operations room never stopped. The headquarters regular working hours were from 08.30 to 18.00 each weekday and 08.30 to 13.00 on Saturdays. Actually, everyone worked as long as whatever was in hand demanded – all staff officers were permanently 'on call'. Like all the other wives, Sue had a part-time husband.

A three-year-old and a one-year-old occupied most of Sue's time. While she was a truly marvellous mother, caring for and entertaining tiny children day after day did not come easily to her. She had always read avidly and craved further intellectual stimulus, which was not available in Londonderry. There was no theatre; we had no access to a cinema; and the internet was still years away. So she set to with making clothes, for herself, the children and friends; she baked, making bread and biscuits for sale in the barracks' Thrift Shop – and always with a full oven, so money was not wasted – and, together, we brewed beer and made wine. I dug a productive veg patch. We kept our fuel bills down by burning 'Sapper logs'. 'Sappers' are Royal Engineers. They built whatever defences the Army needed and took them down when they were no longer required. Amongst the reclaimed material was quite a lot of cheap wood. It was good only for burning and those officers who had open fires in their quarters were able to take it to augment their heating. Soldiers' quarters did not have open fires, so, unfortunately, it was not an option for them.

Everybody was feeling the financial pinch. Army pay was, and still is, determined by the Armed Forces Pay Review Body that had been brought into being in the early 1970s. The government was under an obligation to accept the Review Body's recommendations, but for several years, increases in pay were clawed back by commensurate increases in food and accommodation charges for single soldiers and raised quarter rents for married personnel. The result was that all members of the Armed Forces became poorer and poorer year on year. Soldiers were leaving the Army in ever increasing numbers to find employment elsewhere and it was a similar story for officers. For most officers who elected to stay, it was a case of drawing on savings, or making their insurance schemes paid up, in order to balance their personal finances.

However, it was not all gloom and doom in Ebrington Park. It had its own social life beyond the tea and coffee exchanged daily amongst the wives, but the financial squeeze meant that the highlight of our social life – dinner parties – had to be shared affairs. One family took on responsibility for drinks and the first course; another for the main; and a third for deserts and post meal drinks. Parties in the mess were few and far between and a choice had to be made whether or not to attend them, or play one's part in a domestic dinner party. Both were rarely affordable.

Virtually every household in Ebrington Park had young children. Baby sitters were not easily available, so the answer was baby alarms. During most Saturday evenings wires ran everywhere up and down and across the street. Initially, an alarm would run from a child's home to the first course venue, only to be altered to other houses as the meal progressed. Reeling the wires in on Sunday morning was like a giant game of cats' cradle. And that was not the only dinner party hazard. At some stage, most parties would be interrupted because an incident had occurred and a particular officer, or officers, were required back at duty. A departing host had to trust his mates with his limited supply of booze – not really a risk as the Army's strict limit was a couple of beers, or two glasses of wine a day.

Summer meant leave on 'the mainland' for most people. We combined a visit to Sue's parents with a week spent in north Wales, with Nick and Denise Williams and their children, Simon and Debbie. The four children loved being together – except when Debbie biffed Tim, which she was inclined to do whenever she got over excited. Our holiday home was a somewhat bizarre building, with stained glass windows, a mezzanine floor and a minor labyrinth of corridors and rooms. It made an ideal hide and seek playground for the children. The adults christened it 'the Methodist chapel'. I remember that it rained a lot during the week, but Wimbledon was on, so we watched quite a lot of tennis on a black and white television. Nothing restricted our enjoyment, simply because we were away from Londonderry. We did not have to look under our cars. We could have more than a couple of drinks a day and put rubbish in the dustbin without peering intently at the lid to see if it had been moved. We felt tension ebb away, only to build again as the day of our return to the Province approached.

Leave was to intervene unexpectedly in my life in Londonderry. My colleague, David Daniels, the grade 3 staff officer running the operations room, took his leave and I was told to cover for him while he was away. I moved to take over his desk temporarily. It was situated in a corner of the operations room, which was a large space, perhaps 40 feet long by 18 feet wide. Immediately adjacent to my new desk was a console that ran nearly the width of the room. On it were three or four telephones, plus the hand and headsets for two radio nets. That was where the watchkeepers sat. Beyond them was an open space, until in the corner diagonally opposite to mine, was a desk occupied by the Press Officer. He needed to keep completely up to speed in order to field the stream of media inquiries that constantly came his way. The end wall was given over to a huge map of the brigade's area of responsibility, which corresponded exactly to that of the RUC's Division N. The mirroring of areas of responsibility was important because it meant that 8 Brigade and the local police shared the same geographic focus.

As well as continuing with my own responsibilities, I was now responsible for monitoring the actions of the watchkeepers and keeping abreast of any terrorist incidents as they developed. If units on the ground required additional assistance, I was authorised to grant it, or request it from elsewhere, be it observation helicopters, specialist search teams, a Scenes of Crime Officer from the RUC, the ATO, or dogs. Importantly, I had to keep the brigade commander and senior staff officers continually abreast of any situation in case they wished to intervene.

During the two weeks that I provided cover in the operations room, I worked night on, night off as the on call operations staff officer. I shared this duty with my immediate boss, the Brigade Major. Together we provided a safety net for the watchkeepers, who had us to turn to if a major incident occurred, or they were simply unsure of how to respond to a request or development. Watchkeepers were captains who were with us for six months. They were not accompanied by their families. Invariably, it was the first time they had worked in a headquarters, so their experience on arrival was limited. Normally, one watchkeeper was on duty at a time, but if a major incident occurred and the radio and telephone traffic rapidly increased as a consequence, a second watchkeeper would be called in, as well as the relevant staff officers.

I enjoyed my two weeks in the operations room. It held far more interest than my own job and I was envious of David when he returned from his leave. I was called into my boss's office a day or so later. He explained that I had done well in the operations room and, when David finished his tour of duty in Londonderry, I was to replace him and an incoming officer would take over my current job. I was extremely pleased. Work in the operations room had more consequence and responsibility than my current job and it would help time go more quickly. Despite the importance of the work that we were doing, nobody wanted to stay in Londonderry longer than was necessary.

I quickly slipped back into the routine of my own job, but it was interrupted one day when I felt a pain in my stomach. Over the course of the next few hours the pain intensified and I went to see the Regimental Medical Officer. He felt my stomach and immediately asked his sergeant to telephone the local civilian hospital – the Altnagelvin. I had appendicitis and needed to have an operation as soon as possible.

The Altnagelvin was situated in the east of the city and was not in a dangerous area. However, as a National Health facility, it was open to everyone, making the admittance of any soldier anything but straightforward. Soldiers had to be kept apart from civilian patients and had to have an armed guard with them twenty-four hours a day. I was operated on quickly and efficiently. I woke up in a small side ward with four beds. I was the only patient there. My guard sat outside the door, rifle across his knees. The nurses who looked after me were very pleasant, with the exception of one. She spoke only to tell me what she had to do and while she was efficient, the changing of dressings was done to cause as much discomfort as possible. It was not difficult to work out that she was a republican, possibly with IRA leanings. She clearly viewed me as an enemy, rather than as a patient. I chose not to complain. I did not want to give her that satisfaction.

Towards the end of the year I took over the job in the operations room. Days flew by, but nights were a considerable challenge. Nick, our first son, was an ideal baby. He lay contentedly in the arms of anyone who held him. He did not need cuddling. Right from the start he was a quiet, content, self-contained, little person. He slept through the night, rarely cried and walked at an early age. Tim was very different. He was a people person. He loved being cuddled and the more people who were around him, the better. He delighted in attention and play and could never get enough of it – a big demand on Sue. Left alone he was unhappy. For about two years he could not see the point in walking, when crawling, rolling and bumping around on his bottom was far faster. At night, he woke and cried … and cried … and cried.

I would have a 'night on, night off' duty until the end of my tour. A fortnight of that schedule, while my colleague had been on leave, had been very manageable, but month after month was a different proposition. Sue and I had always shared duties and our children were no exception, assuming that I was in a position to help. Together we worked 'night on, night off' when it came to looking after them. We made my 'child' night coincide with my 'on call' night and that should have worked quite well. However, the majority of 'on call' nights meant that there were, at least, one or two telephone calls for authority to deploy some specialist assistance, or to seek advice on a course of action. Quite frequently I would have to get up and go into work to supervise and assist the watchkeeper as he dealt with a more serious incident. That would leave Sue to look after the children on her supposed night off and that took its toll. We were both drained and increasingly tired and irritable. Strain gradually built on our marriage.

Matters came to a head one evening when I was not as understanding, or responsive, as I should have been, over something neither of us can remember now. We had a heated argument and Sue was utterly distressed. She rushed from the sitting room to lock herself in the downstairs loo. She sobbed that she was leaving. I was aghast. When she finally emerged, for the first and only time in our marriage, I restrained her and held her until we both had recovered a little equilibrium.

I had married a formidably strong person and for Sue to act like that was testimony to the appalling tension and unhappiness that she felt and had kept simmering inside her. We began to talk things through and that took some time – several days perhaps. We were certain of our love for each other and for our children, but the combination of the tensions imposed by the 'Troubles', severe financial concerns, my lack of any reasonable work/life balance – which I admit made me difficult to live with – and her lack of stimulus outside our toddlers, pushed us over the edge that evening. It was the nadir of our marriage. We knew that we had to rebuild. We had the partnership to do it. And we did.

We had to find ways to gain relief from the very restricted life on the Army quarters patch, overlaid by the constant threat from terrorists. There were frequent reminders of how close that threat really was because every bomb that exploded in the city centre, could be clearly heard in Ebrington Park. Additionally, very occasionally, a high velocity round, fired from somewhere on the other side of the River Foyle, would 'crack' over our houses. Both the bombs and the bullets were very disconcerting.

We began our mutual rehabilitation by sharing our darkest moments. Mine was when Tim would not go back to sleep. He invariably woke around 2 or 3 o'clock. As soon as Sue or I entered his room, he would stop crying and happily take a bottle. After that he was awake, wanting company and it would take an age to get him asleep again. When he eventually did, it was possible to tiptoe from the room, but nine times out of ten he would wake almost immediately and, realising that he was alone, would cry again. At its worst this cycle could go on for well over an hour. I was desperate to get some sleep and was so annoyed with my infant son that a couple of times, I had to leave Tim and sit on my hands halfway down the stairs in case I did something that would harm him. I was totally ashamed of my feelings and that increased tension and tiredness. When I confessed this to Sue, she said, 'I've sat on that stair too.' It was an enormous relief to know that it was a problem shared and I was not some kind of monster parent.

We made much greater efforts to make minor expeditions to 'safe areas'. Most memorably we visited the Giant's Causeway; we combined trips to the swimming pool with visits to the helicopter detachment – where Nick was feted by aviator dads who were missing their sons while they served their unaccompanied tours; had walks along deserted, pristine beaches; and one remarkable picnic on cliffs overlooking Magilligan Point. Sue had packed a lunch for us and we left Londonderry one really wet Saturday afternoon. On the cliff top it was blowing so much that the car rocked in the wind. Through rain streaked windows we could see a spectacular seascape and the family was together, enjoying a mini-adventure. Sue broke out the sandwiches, which had been saved from a children's party and put in the freezer. We bit into them. They were still in an almost frozen state – ham and tomato ice lollies will never catch on – and that made the whole thing ridiculous. Giggling together relieved some of the tension and it was an extremely welcome interlude.

Sue also determined that she would not obey all the restrictions on movement quite as closely as she had in the past. She made one shopping trip to Donegal, in the Republic, with some other wives and another outing, on her own, to a shirt factory in Magherafelt, which was on the other side of the Sperrin Mountains from Londonderry. Both expeditions were to look for 'seconds' – products that did not pass the makers' quality control, but their imperfections were so small that they were barely discernible. The Magherafelt trip was a bit of a disaster, because she lost her car keys there. I was rung in the operations room to be told that my wife was stranded in bandit country. I got one of the other grade 3 staff officers to cover for me, signed out my pistol and drove in the operations Mini, with the spare set of keys, to rescue my wife and avoid a considerable embarrassment for us both.

Finding good 'seconds' was a great game and a cost saving measure. Good quality shirts were the favourite commodity, but Tyrone crystal was also high on the list of excellent buys. We also had an unexpected source of low cost food. The Army's headquarters in Northern Ireland was in Lisburn, about 10 miles south of Belfast. That posed a communications problem because the Sperrin Mountains lay between Londonderry and Lisburn and the high ground blocked the signal. Any communications mast, erected in the Sperrins to relay the signal, would have been blown up almost as quickly as it was built and the provision of permanent guards would not be a good use of manpower. The solution was for the signal to be directed to a mast in Scotland and then bounced back to either Londonderry or Lisburn. The mast required routine maintenance and a Puma (RAF troop carrying helicopter), would take signallers from the brigade's signal squadron to Scotland to do the necessary. Wherever that mast was, it was adjacent to a kipper outlet and the signal squadron would take orders for smoked fish, which subsequently arrived by air, at knock-down prices. It's an ill wind that has no benefit, albeit that that was a particularly smelly one.

Something we really missed was eating out. We had not been to a restaurant all the time we had been in Londonderry. Towards the end of our tour, one opened in a large, country house, near Maghera. The restaurant was in the Sperrins, but the house belonged to a loyalist family and was considered to be relatively safe. We decided to treat ourselves. Again I signed out my pistol and a shoulder holster from the armoury and Sue and I set off in our Maxi. We arrived at the house, which was set in its own grounds and had obviously been very grand in its day. It had an air of faded elegance now. The restaurant had been constructed in the cellars and we made our way down steps to be met by a very convivial host. He sat us on a chaise long and, after bringing us the biggest G&Ts we had seen in years, he talked and talked. More than half an hour passed and there was no sign of a menu, but we simply enjoyed the different environment. At last our host suggested that we should eat. He then turned to me and asked if I would like to leave my pistol in his safe! I most certainly was not a James Bond. Not for me a tiny *Walther* PPK pistol and snug shoulder holster – a 9mm *Browning* automatic is a big beast and the service shoulder holster was equally chunky, which meant that my sports jacket was not sufficient to hide it all completely. I pulled my jacket further around me and politely declined his kind offer. It was to be our only restaurant experience while we were in Londonderry, but we had a wonderful meal and a bonus night out.

The biggest operation that I was involved with while serving in Northern Ireland was the Queen's visit to the Province, as part of her tour of the United Kingdom to celebrate her Silver Jubilee. All told, some 32,000 police and soldiers were involved in the associated security operation. The Queen was to conclude her visit with a garden party held in the grounds of the New University at Coleraine and that was on our patch, thus the security responsibility for that event fell to RUC Division N and 8th Infantry Brigade.

For weeks before the visit I was heavily involved in planning. The job of secretary to the RUC/Army Liaison Committee became increasingly important as security measures were refined and adjusted, only to be readjusted as more information became available and the operation took shape. Planning, liaising and writing operations orders probably occupied ninety per cent of my working days during that time.

To ensure the Queen's and the Duke of Edinburgh's safety (and to lighten the security burden), the royal couple were accommodated on HMY *Britannia* for the period of the visit. After a royal reception aboard in Belfast Lough and a land-based investiture at Hillsborough Castle, *Britannia* made her way north to Coleraine. As she entered the waters off the town, she called on the brigade's secure radio net to advise us of her arrival and to come under command for security matters. I had been waiting for that call and was determined that I would be the one to answer it. A voice that could have belonged to a BBC continuity announcer of the 1930s, wafted over the airways –

'Hello Zero [us], this is *Her Majesty's Yacht Britannia*, radio check, over.'

My response was, 'Zero, OK, answer after two seven Delta, over.'

What I had told *Britannia* was that she should answer radio calls after the permanent checkpoint on the Craigavon Bridge! That was lost on the Navy, but all the other users on the radio net knew exactly what it meant and had a chuckle at *Britannia*'s expense.

There had been serious opposition to the Queen's presence in Northern Ireland from republicans across the Province and for days before the visit there had been widespread demonstrations and rioting in most hard areas, including those in Londonderry. That meant me working longer than usual and I was very short of sleep and dog tired by the day of the visit.

As a perk for my hard work, Sue and I were allowed to be present in Coleraine and for me that required a smart uniform. I had only recently been promoted to major, so the Queen's visit was the first time that I had a chance to wear a field officers' hat, which sported a woven gold band on its peak (field officers are majors and lieutenant colonels). Hats were extremely expensive, so while I was at Sandhurst and with splendid forethought, I had bought a secondhand one from a retired Royal Anglian officer. It seemed to fit me quite well. However, the sun shone brightly on the Queen that day, in fact it was a scorcher. The temperature soared and my hat seemed to get tighter and tighter. I came around in a St John's Ambulance tent. I had collapsed briefly, a combination of the hot day, my tight hat and a lack of sleep, so I saw little of the Queen and Prince Philip – but I did get a cup of tea and a delicate sandwich.

That minor mishap could not have mattered less. All the planning worked. The visit was a remarkable success. I got back to the operations room in brigade headquarters later that evening in time to hear HMY *Britannia* take her leave on the brigade radio net as she left our control and laid course for England. Job done.

On the domestic front, there had been a further pay rise for the Armed Forces, but it was another 'give with one hand and take with the other' affair. While things had been very difficult up to then, they were critical now, particularly for married soldiers. Two things happened in 8 Brigade that illustrated just how badly soldiers were suffering.

1st Battalion, The King's Regiment had been replaced as the resident battalion by 2nd Battalion, Coldstream Guards. Aware of some severe dissatisfaction, the commanding officer conducted a survey of the financial well-being of married guardsmen (private soldiers). The results were compelling. After reinforcing a unit responsible for maintaining order in the Bogside, a company of Coldstreamers returned to Ebrington Barracks and its married soldiers went back to their quarters. Some found that their electricity and gas had been cut off by the utility companies. Married private soldiers, with children, simply could not afford to pay their way, notwithstanding that they were managing their income carefully. The rub was that in the Bogside and Creggan, the lights were on and the gas burned brightly, while many of the inhabitants had no intention of paying their bills. To have cut off their utilities was deemed to be inhumane, but apparently it was quite acceptable to deny those same essentials to people who risked their lives trying to keep the peace. The case was sent up through the chain of command in the hope that it would do some good. No-one held their breath.

The second telling event involved an unmarried private soldier of 1st Battalion, The Royal Hampshire Regiment, another resident battalion based in Ballykelly. After he received his pay rise, he wrote a letter to one of the 'Red

Tops' claiming that the cost of the notepaper, envelope and stamp that he had used, accounted for all of the recent increase in his weekly pay. The 'Red Top' checked the claim and it was well founded. The paper ran a campaign and, at last, there was some public awareness of the problem.

Time was almost up for us in Londonderry, but there were a couple of highlights left. One was our leaving party. Friends arranged it, not in the mess, but at an old Martello tower at the end of the Magilligan promontory. The tower guarded the entrance to Lough Foyle at its narrowest point. It was separated from the Republic of Ireland's shore by little more than a mile. It was a really spectacular venue. I had a particular responsibility for liaison with the helicopter detachment and my family was well known to them from my weekend visits with Nick, so I approached the detachment commander to see if it would be possible to get Sue a ride in a *Gazelle* reconnaissance helicopter. He agreed, provided that it could be integrated with a routine patrol – even that would get him into serious trouble these days. On the evening of the party I delivered Sue to the helicopter pad in Ebrington Barracks and left her there to await her lift, while I drove to Magilligan. I was about half way there, driving on the coast road, when I saw the *Gazelle* about 800 yards from me following the water's edge. It was nose down, less than five feet off the sand and going as fast as it could. As it approached the Martello tower, what I did not see was the pilot pull up steeply to several hundreds of feet, to execute a stomach churning stall turn and land amongst the sand dunes. Sue had had the taxi ride of her life and, even today, she smiles at the memory.

As at Sandhurst, we were kindly given a splendid last night's dinner – this time our hosts were the Brigade Major and his wife. Ebrington Barracks had previously accommodated the Royal Navy – Londonderry was the principal base for convoy escorts during World War II. The Brigade Major's quarter was formerly the naval Captain's house, which was situated to overlook the River Foyle. It had an impressive dining room and we were on the point of finishing our meal when there was an enormous explosion, which rattled the windows. We all ran to pull back the curtains and were in time to see pieces of debris fall into the river from the main shopping area on the opposite bank.

'Were we expecting that?' the Brigade Major asked.

'Don't think so,' I replied – often we would get incomplete intelligence that the IRA had something in the offing, without any specifics attached.

'Are you on duty, or am I?' my boss said hopefully.

'Give you one guess and it's not me!' I said firmly.

Londonderry was over. True to form, it went out with a bang.

During the writing of this book I have made frequent use of Sue's guest book, which she has maintained since we were married. It has helped enormously to confirm dates and places. Her entry for Londonderry is 'No one came to stay!' Only five words, but they speak volumes.

CHAPTER ELEVEN
STAFF COLLEGE – FROM EXTRA MATHS TO A COAT OF MANY COLOURS

IN OCTOBER 1977 the family moved to Shrivenham in Wiltshire, to the Royal College of Military Science (RMCS), which was to be my first place of study. On selection to attend the staff college course, officers were allocated to one of three divisions. Divisions 1 and 2 were for those with a scientific bent and their courses at Shrivenham lasted for a year. They were destined to work on the technical staff, involved in the design and delivery of new weapons and equipment. I was only ever a contender for Division 3, which involved a ten-week briefing and study of technical advancements, and the development of weaponry and other equipment that an army needs to function efficiently.

While we were still in Londonderry, I had received news that the Pompadours next posting was to be in Belfast as a resident battalion. I immediately realised that because I had served two consecutive tours of duty away from commanding soldiers, I would be required to return to regimental duty when the staff course was complete. I thought it would be grossly unfair on Sue to have to suffer another extended stay in Northern Ireland and I was not much attracted by the resident battalion role, or the 'rent a company' concept that went with it. I put my case to change battalions to Regimental Headquarters and it succeeded. I was rewarded with a posting to the 2nd Battalion (the Poachers) once our antipodean adventure was over. They would be in … BERLIN. It was difficult to stop smiling.

While I settled down to read the impressive number of précises that would get me through the first week at RMCS, Sue had a more difficult job to address – the onward move to Australia. It was a logistical nightmare, conducted in the smallest, most cramped quarter that we had ever lived in. She had to devise a three-way split of our possessions. Our furniture would be put in store until we returned to the United Kingdom at some unknown time in the future. Stuff we needed in Australia had to be identified, packed into MFO boxes ready for shipment there and, to complicate matters, we were restricted in the amount that we could take. Finally, the residue of our belongings were to be consigned to yet more, separately marked, MFO boxes, initially destined for store, before being forwarded to Berlin in about twelve months' time. She did it brilliantly.

I joined my syndicate (from memory, about eight or ten officers) and the course began. I found it all hard going and time consuming. Even with pre-reading, lessons were a severe trial, although I probably grasped most of the essentials. Maths was my real *bête noire*. Three attempts to obtain a pass in the subject at GCE level was no preparation for what confronted me now. I was completely lost and I was not alone. Dair Farrar-Hockley was a Parachute Regiment officer and he was as clueless about maths as I was. Our syndicate officer kindly arranged for a civilian lecturer to give us personal instruction at the end of the working day. A chap of about our own age (mid-thirties) turned up and was entirely confident that he could provide us with all the assistance we needed. He said that he would start right at the beginning and proceeded to write a series of numbers, letters and symbols on a blackboard. For all I knew, I could have been looking at the chemical formula for Fairy Liquid! I turned to see Dair looking at me. He shrugged his shoulders and mouthed 'What?' The lecturer caught the moment. 'You don't understand at all, do you?' he asked. We admitted our shortcoming. He looked at his watch – 'The bar's open. The best thing we can do is go and have a drink' – end of extra maths.

Despite my difficulty comprehending some aspects of the instruction, the course was very interesting. Real experts, who were enthusiastic about their subjects, delivered the central lectures. The practical work in laboratories and workshops was often entertaining. One week was given over to ballistics and we were taken to a room that had a powerful machine that fired tennis balls at huge velocity. The exercise was to calculate the speed of the projectile and photograph it, in flight, at a particular point, with a specialist camera. To make it more difficult, the camera was positioned at right angles to the tennis ball's path. My slide rule did not help a bit. After a considerable number of attempts I got a clearly defined photo, but the shame was that there was no sign of the tennis ball. What I had captured was a three-point electrical socket on the wall opposite the camera. I was not weapons staff material.

There was no trooping flight to Australia and so the family would travel directly courtesy of British Airways. Sue did some research and found that for exactly the same price, we could go by Thai Air and have a three-day stopover in Bangkok that offered the double advantages of a break in a long flight and it would allow us to see something of a country that we had not visited before. I had to apply to a MOD department for the British Airways flight tickets and asked if I could be advanced the money to purchase the alternative flights with Thai Air. The answer was a resounding 'No'. The solution was a bridging loan, kindly afforded by my mother. I bought the tickets and submitted my claim for reimbursement, which was honoured in less than twenty-four hours – a case of bureaucracy gone mad.

The students' major project at RMCS was to write a technical paper on a subject of their choice. I chose clothing, because, although it was improving, there remained a widely held view that the Army's combat clothing remained inadequate for the environments in which soldiers had to operate. The wet cold that persisted for much of the year in Northern Ireland, comprehensively defeated the then combat kit. The combat smock and trousers were far from waterproof: the leather of the ankle boots was little better than compressed cardboard and, quite literally, could not keep out a heavy dew: and the first attempt at a combat glove was an improvement on its woollen forerunner, but it still had a long way to go. The combination of all that meant that soldiers who, in rural areas, conducted patrols of several days duration, often returned to barracks with varying degrees of foot rot and hypothermia. Many resorted to buying kit from outdoor pursuits' manufacturers in order to stay warmer and drier than they otherwise would.

Wet proofing was the focus of my paper and I arranged a visit to the Stores and Clothing Research Establishment at Colchester to confront the experts and ask why they had allowed woefully poor clothing to be introduced into service. Once there I was entirely disarmed. The scientists welcomed me very warmly indeed. Far from being defensive, they wanted to hear my criticisms and to discuss them with me. Then it was their turn. They were keen to show me what they were working on and their projects were wide ranging, from single fibre filling and reduced cold spots for the next generation of sleeping bags; to a combat glove that promised to keep hands warm and dry; to wet proof smocks and trousers that, I believe, were the forerunners of the Gortex and other famous makers' range of outdoor clothing. They seemed to have an answer for everything, except for that which concerned money. In essence, their simple message was that inadequate budgets meant inadequate kit. Fortunately, the late 1970s saw a turning point. The cold, wet, Northern Irish climate had such a debilitating effect on soldiers' performance that things had to change. Since that time the British Army's clothing and other personal equipments have improved out of all recognition.

As time at Shrivenham drew towards its end, the move to Australia dominated our thinking. Using Sue's three-way split of our possessions, we began to pack our MFO boxes destined for Australia – they would be despatched first in order that we would not have too long to wait for them after their journey by sea. Then we had a stroke of luck. I was advised to contact RAF Lyneham – only a short distance from Shrivenham – to see if there were any military cargo flights to Australia in December. There was one and it had spare capacity on the outward leg. It would take our boxes and they would be waiting for us on our arrival.

The next conundrum to be addressed was the selling of our car and white goods (the latter do not do well in store). We needed all of them almost up to the date that we left our quarter, but they had to be gone by then – tricky timing, but we achieved it after some considerable effort and negotiation. All that was left then was to clean the quarter and hire a car to transport the four of us to Sue's parent's home for Christmas.

Festivities over, Sue's father delivered us to Nuneaton railway station on a cold, blustery, New Year's Eve morning. A bitter wind swept the platforms. I looked at my family huddled in a tiny group, swaddled in anoraks and scarves, surrounded by baggage and thought of the enormous change that awaited us.

We flew on New Year's Day. The Thai Air staff were marvellous. As a young family we boarded the aircraft first and were given a row of seats, with masses of leg room, immediately in front of a large television screen. The hostesses loved Nick and Tim immediately and even before we took off, the boys had Thai Air Christmas stockings. During the flight to Bangkok we crossed the date line again and more presents were showered on our sons. Thai Air's hospitality was impeccable, making the long flight as pleasant as possible for all of us.

The aircraft arrived in Bangkok in the early hours of the morning. A taxi transferred us to our hotel through near deserted streets, but the sights, sounds and smells were very different from anything that we had experienced before. Our first priority was sleep. While Sue did whatever unpacking was necessary, I took the boys for a shower. Tim was soaped all over and sprayed off and then it was Nick's turn. Immediately, I noticed a pustule in the centre of his back. He had chicken pox! Kids' sense of timing never ceases to amaze and complicate matters.

Despite the spot on his back, Nick appeared fine and we embarked on the first of several tours of the city to take in the usual tourist attractions. Our first outing was in a people carrier that had an Italian civil airline pilot and his dramatically elegant wife as our fellow passengers. From the moment we got into the carrier, they made it abundantly clear that they thought travelling with a young, noisy English family was unacceptable. However, fate played its part and our acquaintanceship was cut short when Tim projectile vomited across the back seat. Magicians could not have conjured a disappearance as instantaneous as that made by the Italian couple. I was so pleased with my son. The taxi driver was not in the least disconcerted and took us to buy Tim a new set of clothes, while he cleaned the vehicle. By the time we left Bangkok we had gorged ourselves on temples, palaces, dance troupes, river trips and floating markets, but the experiences had been well worth the battle with the MOD's 'job's worths'.

The flight to Sydney was almost as long as that to Bangkok. We landed at about 10 o'clock at night and the aircraft was taxied to a remote part of the airfield. Two Aussie immigration officers came on board and used huge aerosols to spray everyone liberally before the aircraft was allowed to taxi to its stand. Sue and I were disconcerted by the stringency of the Australian arrival procedure, not least because Nick had developed a few more spots, including several on his face. We thought that we had better cover those, so put him in his anorak, with the hood up, despite the temperature being well into the twenties centigrade. I presented Sue's passport to an immigration officer (the boys were included on hers) and gave him an Australian Army document that identified me and gave me and my family the right to stay in Australia for a year. He had not seen one before and it took quite some time for him to accept my explanation. Then he focused on Nick. 'Why is he wearing an anorak?' It was an obvious question to ask. 'He's upset at leaving UK and the anorak is his comfort blanket,' was the best we could manage – hardly convincing, but he let us through.

Our trials were not over. We held a reservation for a family room in a motel, not far from the airport. It was now nearing midnight and we were all looking forward to a decent sleep. The motel's reception area was deserted, but there was a lot of noise coming from the bar. I opened the door on a room full of men having a really good night out. No one paid the least bit of attention to us. I approached a bar tender and showed him my reservation ticket. He went to a book by his till.

'You're not booked in here, mate,' he said and went back to serving drinks.

I broke the news to Sue and she told me that it was my job to sort it out. I returned to the bar, located the manager, but he was of no help. An Aussie customer, with a beer in his hand, asked me why I was in Australia. I explained as quickly as I could.

'He's in the bloody Army. Give him a room,' was the very welcome support I got from my newfound friend.

'Why didn't you say?' said the manager and the problem was solved – or nearly. 'Put your wife and kids in the room and come and have a drink.'

It was clearly more of an order than an invitation from the manager and a couple of hours later, tired and seriously woozy, I finally got to bed.

Next day we flew to Melbourne to be met by Lieutenant Colonel Nigel Still of the 17th/21st Lancers. Nigel and another British lieutenant colonel, Compton Boyd, were serving two-year exchange postings at the Australian Staff College as members of the directing staff. Nigel had transport arranged to take us to the Travel Lodge in Geelong, where we would stay for several days until our hired bungalow became vacant – a hiring because the Australian Army did not have quarters in Queenscliff, where the Staff College was located and all married officers' accommodation was rented from the civilian housing market. Prior to leaving England I had contacted my predecessor, an officer in the Royal Artillery, for advice on obtaining a hiring. He happened to be married to an Australian lady, whose family owned a property in Point Lonsdale, a small seaside town only a couple of miles from Queenscliff. He and his wife had lived in it while he attended the staff course and he advised that it would be ideal for us, so that was the accommodation problem solved.

Our next hurdle was to buy a car and again my predecessor was helpful in providing a contact. The car guru was a Corporal Sarrich of the Royal Australian Electrical and Mechanical Engineers, who worked as a vehicle mechanic at the Staff College. I wrote to him, while still in the UK and we agreed a plan. The arrangement worked well. Corporal Sarrich contacted me, established what my budget was and then left the matter there. The following day the telephone rang in our hotel bedroom at well before six in the morning.

'Corporal Sarrich here, sir. *The Age* has just hit the streets and the Saturday edition has a big used car section. I'm in Melbourne [50 miles from Geelong] and I've looked at three cars. I've got the one for you. I'm coming to pick you up'.

I was staggered. Heaven knows what the first two car owners thought when they were woken at some ungodly hour, to have their cars inspected and then rejected. By 9 o'clock I was the owner of a sun-faded Ford Falcon straight six [cylinders], with a bench seat, three forward gears and a column shift gear change.

On the way to Melbourne Corporal Sarrich was insistent that I pay by cheque. He told me that at some time in the year, the car's engine bloc would have to be replaced, but that that would be cheap to do and I would recover my money at the end of the year. I paid Aus$800 for the car and was relieved to get back to Geelong, having successfully negotiated Melbourne's traffic in an unfamiliar vehicle. I was not allowed to rest for long. Corporal Sarrich rang again.

'I've told the [previous] owner that the car broke down on the way to Geelong; you've cancelled the cheque and are pissed off that someone has taken advantage of you when you had only just got here.'

I protested that it was not true. Corporal Sarrich assured me that Aus$600 was what the car was worth and he had already negotiated that revised sum with the owner, claiming that Aus $200 was needed to repair the vehicle. I signed a new cheque and the initial one was recovered. As the year progressed, I was to learn that when it came to cars and deals, Corporal Sarrich was a force of nature and, on leaving, I recovered my outlay as he had promised.

We moved to our new home at 3 Alexander Crescent, Point Lonsdale on a particularly hot day in January – high summer in Australia. Our first impression was very favourable. It was a colonial style bungalow with steps to a white-painted front veranda; a driveway led to a large garage; and the property had pleasant, established gardens front and rear. Inside it was a different matter. The front door opened onto a spacious lounge. At one side of the building were two bedrooms and the bathroom. Those rooms were what estate agents might term 'compact'. What became the children's bedroom, had a curtain across its entrance rather than a door. On the other side of the building was an adequate kitchen and built onto the back of the house was an enclosed veranda that served as both our dining room and the children's playroom. Tacked onto that was a 'jerry built' toilet that had no electric light, or proper foundations. Finally, in the middle of the building, enclosed by the lounge, kitchen, rear veranda and children's bedroom, was another small room that would become my study. All the furniture was old and tatty, the sort of stuff that had passed its 'sell by' date in the owners own home, to be moved to the holiday property as a matter of convenience. Unfortunately, we had taken my predecessor at his word and had already signed a year's contract for the bungalow. After all, we thought this place was 'ideal'. At that point we had no way of knowing anything different, but a much more immediate domestic crisis occupied our attentions.

Nick's chicken pox was running its course without any undue alarms, but Tim had contracted the illness too. Very quickly he was covered in spots and as the days passed, his spots turned to blisters that were everywhere over his small body: between his toes, in his ears, on the inside of his nose and clusters formed around all the damp areas of his body. He was dreadfully distressed, seriously uncomfortable and very ill. We had taken him to a local GP soon after he had contracted chicken pox and the doctor was particularly concerned that Tim should be kept out of the sun. He explained that exposure to sunlight could leave Tim pockmarked for life and he should be kept in dark rooms as much as possible. During sweltering summer weather, the poor little mite was virtually confined to his bedroom for the next couple of weeks and was slathered in calamine lotion at least twice daily – however, he recovered well. Not much has ever kept his bright disposition down for very long.

While Sue contended with our sick sons, I had to become acquainted with the Staff College and Queenscliff, which was then a small, sleepy holiday resort at the end of the Bellarine Peninsular. The College was housed in a brick built fort, formerly home to a coastal artillery battery. The gun emplacements were strategically placed to look out over a spectacular seascape, to command the narrow entrance to Port Phillip Bay, which gives access to Melbourne. While the College did not have the grandeur of the British Staff College at Camberley, it was an attractive, homely place and the Australians were rightly proud of it.

The thirteen foreign students formed up early for a series of orientation briefings. Two of us came from the UK. Others were from the USA, Canada, New Zealand, Papua New Guinea, Indonesia, Thailand, India and Pakistan.

My opposite number was Major Tony Redwood-Davies from the Duke of Wellington's Regiment, another infantryman. He was a very capable officer, who enjoyed a good time and we got on well.

After one of the early briefings we both needed to go to the loo. We found one, a tiny brick built construction with a wriggly tin roof. It would not have been out of place on a building site. Directly in front of us, on the wall above a long, rudimentary urinal, was a brass plaque. Its text informed us that we were in a listed building! Apparently, anything built in Australia before about 1920 qualified for listing. We were far from sure about that and given the building was a loo, we were certain that the Aussies were taking the ... well you know what I mean!

The main body of Australian students gathered at the College a few days later, bringing the student body to a total of eighty. The Australian Army is small and it was immediately apparent that the Aussies either knew each other personally, or that they knew of each other, even if they had not met until that point. It generated a family atmosphere from the outset. This closeness was enhanced by the formal appointment of the 'senior student', Peter Schuman, a Special Air Service soldier. His job was to provide a bridge between the student body and the directing staff and it was quickly apparent that he had a magic touch. Throughout the course he held a briefing before work each morning, passing on instructions from the directing staff and gathering feedback from the students. Peter was able to 'oil the wheels' and within a very few weeks, the directing staff let it be known that they considered us to be an able bunch and, if we were reasonably industrious, they would do all they could to make 1978 a year to enjoy.

The inevitable 'get to know you' party was held and as the party progressed I found myself next to an officer called Bob Carson. Bob had spent 1977 at Shrivenham as a Division 1 or 2 student (the Australian Army did not have an equivalent to RMCS and every year it sent several officers to receive technical and scientific training in Britain). I had not met Bob during our three-month overlap at Shrivenham, but he knew very well the financial straits that all Brit soldiers were in.

He took me to one side to say, 'I'll know you'll be broke, mate, but I want you to enjoy your year here. If you ever need a loan to go and do something, don't be shy to ask.'

It was an amazingly generous offer, if somewhat embarrassing. It also was typical of the kindness the Aussies showed to their guests throughout the year.

Just before the course began we had bad news. I was informed that my MFO boxes were in a customs store at the Royal Australian Air Force (RAAF) base at Edinburgh, near Melbourne. I had been sent various customs forms while still in England and had filled them out assiduously and returned them. Clutching duplicate forms I set off in an Aussie Army lorry to collect my goods. I found the store and could see my boxes piled in a locked cage. I spoke to an obdurate customs official only to be told that I was late; that the store only opened a few times a year and my stuff was impounded until mid-March. I hurried back to the Staff College, but the authorities there could only apologise for not notifying me of the customs' opening times. Sue was less than amused and Nick and Tim were upset because Santa had put all their big Christmas presents in the MFO boxes to give his reindeer a break.

After that we needed some good news and it came when Tony and I were called to the British High Commission in Canberra. There we were briefed by the military attaché, a full colonel in his fifties. His principal objective was to ensure that we did not transfer to the Australian Army at the end of the staff course. Apparently, all of the last eight or so British officers, who had served on various exchange postings in Australia, had transferred, because the terms of service offered were so much better than those of our own army. With years of British training behind them, they were welcomed with open arms. The colonel concluded his briefing by confiding that he was married to an Australian lady and he too would be retiring and staying in Australia. He told us that he had gained a good understanding of Australia and we would be wise to regard the country as somewhere like Nigeria – not a place just south of the Cornish coast, where the Queen reigned, cricket was played and people formed orderly queues. Tony and I looked at each other. Had the poor old boy lost his marbles? The colonel picked up on the glance.

'What I mean,' he said firmly, 'is that at some time in your year, the Aussies will behave utterly differently than Brits and it will disappoint you if you aren't expecting it.'

We nodded an understanding, but were less than convinced.

Our next port of call in the Commission was the pay office. There a sergeant in the Royal Army Pay Corps greeted us by saying that he was there to help and any financial worries we might have would soon be mitigated. Manna from heaven! At that point I had served fourteen years in the Army; I was thirty-four years old; a major and probably in the top ten per cent of my peer group; I was not in debt, but despite all that I had only £30 in readies to my name.

The sergeant produced two sets of forms already filled in.

'This one grants you one month's advance of pay and the second one gets you another. You will have to pay it all back over twelve months, but it should help with set up costs and leave you with some money to spare.'

He went on to explain that two obscure paragraphs in the pay manual permitted the advances. We signed, thanked him profusely, took the money and ran.

The Staff College authorities had done all they could to support foreign students. Each one of us had a student, a directing staff and a civilian sponsor to help us with various aspects of work and social life. I owe a lot to them, but particularly my student and directing staff sponsors. The student sponsor was Major Kevin Fletcher, an Armoured Corps officer. Brits might think of him as the archetypal, dinkum Aussie, although he would cringe at the description. He helped me deal with day-to-day occurrences that would be new to me, simply because I was in a foreign country, serving in a different army. My directing staff sponsor was Lieutenant Colonel Graham Burgess, a kind and experienced artillery man. He gave advice on the course and how to approach the various areas of study.

The course got underway. It had three terms, with very little break between them. The student body was divided into syndicates, each of about eight officers, with a syndicate leader from the directing staff, mirroring the arrangements at Shrivenham. Every student had his own pigeon hole that required clearing every day. The administrative staff would place précis and other materials there that had to be read in preparation for forthcoming instruction and projects. The pigeon holes became the bane of students' lives for however hard you tried to keep on top of the reams of paper placed in them, there was always something else waiting when you looked again.

I quickly realised why it was important to have foreign students on the course. The American, Indian and Pakistani armies are vast and students from those countries were familiar with manoeuvring and supporting large numbers of troops. While the British Army was smaller, in 1978 it could still field a Corps (over 100,000 troops) and its technology was second only to that of the Americans. Together, we foreigners introduced different perspectives to those common amongst our Australian colleagues and our opinions were valued because of it.

Apart from the mound of reading to assimilate, there was a pile of written work. Before we embarked on our first papers, a series of lectures was delivered by Lieutenant Colonel David Cross of the Royal Australian Army Education Corps. He was an outstanding speaker and his subject was the written word. His aim was to improve our writing abilities. One piece of his advice remains with me – draft your work as well as you can; if possible, put it aside for some time; then reduce it by ten per cent in order to remove the flab, tighten the prose and heighten the meaning. He warned that a reduction of ten per cent would rarely be achievable, but any saving would help.

At home Nick and Tim had recovered and we were settling into our new environment. What we first thought were rats in the roof of our bungalow, turned out to be possums; we had got used to the night-long croaking of frogs mating in the front garden's pond; we were learning the names of the various exotic plants in our generous gardens, where the boys played endlessly and, for them, permission to wee on the lemon tree was a 'boy joy': but conversely, both of them had been made aware that crawling under the bungalow's raised foundations – it was built on a slope – was absolutely forbidden because of the snakes, spiders and heaven knows what else that resided there. Nick was signed up for the local kindergarten; Sue was playing tennis and making friends amongst the other wives; and we had mastered the idiosyncratic column gear shift on the Ford Falcon. We acquired a few essential items of furniture to augment the bungalow's meagre inventory and with those installed we were all set for the year ahead.

The Australian Army did not stint on the staff course. Probably, once in every month the course left Queenscliff for two or three days, or longer, to visit various schools and units of the Army, Navy and Air Force. It was a costly exercise, but the visits were extremely valuable professionally. One of the early visits was to Sydney, when Australia experienced its first terrorist attack by a member of the Ananda Marga organisation, with the bombing of the Sydney Hilton Hotel. This was a particularly sensitive target because the hotel was playing host to a Commonwealth Heads of Government regional meeting at the time. Sadly, a policeman and two refuse workers died and the Australian Prime Minister called on the Army to take on the security responsibilities. As a direct consequence, counter terrorism was rapidly promoted in Australian military thinking and my experience in Aden and in Northern Ireland was in demand.

Most Aussie officers worked hard and virtually all of them played hard. Babysitting circles were formed to free up attendees for a constant stream of weekend drinks and BBQ parties and to allow the wives to follow daytime pursuits like tennis, book clubs and bridge. It was our babysitting circle that first alerted us to the fact that in hiring our bungalow, we had been duped. Normally, it was the men who did the babysitting because it allowed us to get on with our staff college homework in comparative peace and quiet. After my first stint, I reported back to Sue that the house I had just been in was infinitely smarter than ours. My second round of babysitting confirmed our suspicions – another pleasant, modern house – but we were stuck with the place we had got and so made the best of it.

Towards the end of the first term some of the students and wives got together to form an am-dram group and they rehearsed to put on an evening of short, one act, Victorian melodrama plays in the Staff College's dining room. The rest of us were invited to provide the audience and to dine. Not content with that the organisers wanted music between courses, so a barbers' shop group was formed to sing half a dozen songs. I was pressed ganged into joining the barbers' shop group and the rehearsals and production turned out to be enormous fun.

The end of term approached and Sue and I thought it would be good to use the few free days allowed to us to visit Tasmania. Ever keen to help, a fellow student called Chris Welburn insisted that we stay the first night with his sister's family in Devonport. We flew from Melbourne in a small aircraft and landed at an airport, some miles outside Devonport, at around 9 o'clock in the evening. It had little more than a big shed as its terminal. We had hired a Mini, but somehow two cars turned up, both from the same company. We left the hire car representatives to sort out the muddle and drove to the address that we had been given in Devonport.

It was now dark and the town was quiet – that is all except the house that was to be our resting place for the night. As I got out of the Mini, I could hear the mother and father of all domestic arguments coming from the house. We had no other accommodation arranged, so, with some trepidation, I knocked at the front door. Footsteps approached and the door opened to reveal an angry looking man who was short on pleasantries.

'Yeah,' was all he said.

I said, 'I'm Colin Groves,' expecting my name to be familiar to him.

'So,' was his response.

I explained the arrangement that I thought had been put in place between his wife and her brother. That relit the blue touch paper. His wife claimed no knowledge of us and she was carrying a new born baby as she came to the door. The argument between husband and wife flared to reach a new level of intensity. I quickly apologised for the misunderstanding, made a hurried exit to the car and an even faster getaway. We had seen a couple of motels on the way into Devonport and were lucky enough to secure a comfortable family room for the rest of the night.

Tasmania was supposed to be a lot like England, but the similarity was difficult to discern. Many of its roads cut through primeval forest; in agricultural areas, there was a distinct lack of hedges; the rural towns still had a whiff of the frontier about them, but we did experience some familiar thick fog in Hobart and names like Sheffield, Bridgewater, Swansea and Westbury let us know where the island's heritage lay.

Anzac Day, 25 April, has special significance in the Australian and New Zealand calendars. It celebrates the exploits of the Australian and New Zealand Army Corps at Gallipoli in World War I. The Aussies and Kiwis were under British command and their casualties were horrendous. Notwithstanding that, the towns on the Bellarine Peninsular always wanted to have a British officer from the Staff College speak at their commemorations. I was sent to Drysdale, a small, rural town and was met by a formidable array of dignitaries. A procession formed and I was embarrassed to be positioned immediately behind the band and in front of everyone else. We set off to a noise like tortured cats. The band were giving it all they had got. It was anything but musical, but good for them. We left the urban area to climb a hill to the cemetery and I was surprised to see the Union Jack flying above the burial plots. I delivered my speech to polite, muted applause. I looked down and at my feet was a commemorative stone to a soldier of the 56th Regiment of Foot, who had served in the Maori Wars of the mid-nineteenth century, and who, on his discharge, had been cared for in a mini version of the Royal Hospital, Chelsea, established with funds raised by the local Drysdale community. He was a soldier of a forerunner regiment to my own. I imagined that if the deceased soldier could have spoken, it would be, 'About time an officer turned up!'

After the ceremony we retraced our steps to the town and to a reception. It was a marvellous community celebration. The afternoon wore on and I attempted to make my excuses. 'Oh no, you're staying here tonight,' was the response I got. I woke up in an unfamiliar bedroom the following day – heaven knows what time it was. My uniform was laid out, pressed, on a chest of drawers. I got up, showered and still unshaven, emerged into someone's dining room. Several people were picking at a meal. They cheered at my appearance – and we all immediately regretted the noise. I had the most immense hangover. It was one amongst many. Eventually, I took my leave and drove slowly home. I had missed a day's instruction at the Staff College; nobody expected anything less.

The second term progressed and the much anticipated, round Australia, study tours loomed large. The student and directing staff bodies were split into two so that one group would go westabout the country in June, and eastabout a month later. The other group had the same itinerary, but in reverse. As excitement mounted some of the Aussie officers were a little worried because at some points on the tours we would be over 1,000 miles from the nearest brewery. Such was the excellence of our planning staff that one of the trucks that carried our baggage to the departure airfield, was devoted solely to the liquid amber! When it came to loading the baggage on the 'A' model (very old) C-130 Hercules aircraft for my westabout tour, the beer had to be stored on the back ramp, but there was so much of it that the upper and lower ramps would not close completely. First of all I was amazed that alcohol was allowed on a service aircraft and I was then astounded when the staff sergeant air quartermaster produced a long, webbing strap, passed it around the outside of the rear ramp, tightened and secured it, in order that we could fly for the next day and a half, at a low altitude, with daylight pouring through the gap at the rear of the aircraft, until we had drunk enough for the doors to close. That certainly was not the way things were done at RAF Lyneham!

The extent of the tours was astonishing. They were designed to give the Australian students first-hand experience of strategic targets (mainly vast mining operations) that could attract the unwanted attention of any foreign adversary and the vital infrastructure that would have to be secured to support significant military operations. Both tours brought home just how big the Australian continent is, with much of it being hostile and uninviting. The Australians called it 'GAFA' – Great Australian Fuck All – and it summed up the interior's red desolation perfectly. The itineraries were:

WESTABOUT:

Depart Melbourne west for **Kalgoorlie** – goldfields.

Kalgoorlie south west to **Albany** – the oldest town in Western Australia.

Albany north west to **Perth** – Western Australia's state capital and the most isolated capital in the world (2,736km east to Adelaide and 3,766km north to Singapore). A major cultural and industrial centre, with a port in nearby Freemantle.

Perth north to **Learmonth** – to visit a huge communications station run jointly by the US and Australian Navies, situated on the North West Cape (back of beyond).

Learmonth north east to **Port Hedland** – important port for the export of minerals, iron ore, salt and manganese.

Port Hedland south to **Mount Newman** – vast open cast iron ore mining operation.

Mount Hedland north east to **Broome** – old port town formerly known for pearl shell fishing, latterly a tourist centre.

Broome north east to **Kunnunurra** – on the border between Western Australia and the Northern Territory. Centre for the Ord River irrigation system.

Kunnunurra north east to **Tindal** – I have no recollection of it. Now a major RAAF station. Perhaps it was then.

Tindal south to **Alice Springs** – central communications hub connecting Australia's interior to the wider world. Fruit and cattle farming.

Alice Springs south east to **Melbourne** – diverting to make a low pass over Ayers Rock (Uluru).

Total distance by air – 10,500km.

EASTABOUT:

Depart Melbourne north east to **Brisbane** – State capital of Queensland and major port city exporting a wide range of minerals and agricultural and engineering products. Large Air Force base nearby.

Brisbane north west to **Mackay** – Tropical port city with specially constructed harbour for the export of sugar cane. Exports augmented by the development of massive open cast coal mines.

Mackay north west to **Townsville** – Second city of Queensland. Major industrial centre and port for the export of minerals from Mount Isa. Significant military presence.

Townsville north to **Cairns** – natural harbour. Sugar exports.

Cairns north west to **Nhulunbuy** (Gove) in the Northern Territory. Aluminium deposits.

Nhulunbuy west to **Darwin** – administrative centre for the Northern Territory. Then the only major port accessible by road on the north Australian coast.

Darwin south east to **Mount Isa** – Back into Queensland. Uranium oxide and copper mining – in the late 1970s 2.75 million tonnes of ore and waste mined annually.

Mount Isa south to **Broken Hill** – New South Wales. Silver and lead deposits. Darling River water pipeline.

Broken Hill south east to **Melbourne**.

Total distance by air – 7,800km

The tours form one of my outstanding lifetime experiences, privately unaffordable for all but the very rich. However, it was on these tours that the British military attaché's prediction was realised. From the time we left Melbourne until we returned, on both tours, there were perhaps three or four Australian students and one or two Australian directing staff who were never properly sober. None of the foreign students were impressed. The behaviour of those Aussie officers was unacceptable in our book though nothing was said.

One incident went down in the folklore of the Class of '78. We were visiting RAAF Amberley, Australia's largest air force base, south of Brisbane. The base operated F-111 swing-wing fighter bombers and the RAAF were rightly proud of them and wanted to demonstrate them to us. We arrived at station headquarters and were ushered into an impressive rotunda with a mezzanine floor above us. The station commander appeared and addressed us from the mezzanine. Suddenly, in the centre of the student body, one of those who was worse for wear splattered vomit onto the floor. There was an embarrassed silence, then 'Bloody good job it's a mosaic!' was the only and very apt, antipodean comment.

Sue was not content to be left out from seeing as much of Australia as possible. She and the wife of the Canadian student, Irene Jacubow, decided to take a two-week bus tour into the interior. It was an epic journey of over 7,000km. Their transport was an old, rattling, outback bus, so their journey was hot, dusty and bumpy. Apart from a night spent in underground accommodation at Coober Pedy (opal mines), they enjoyed rudimentary camping for the rest of their tour. They traced a huge circular route taking in Bourke in New South Wales, then north as far as Cloncurry and Tenant Creek in the Northern Territory, next south to Alice Springs and Uluru, into South Australia and Woomera, before bypassing Adelaide and following the Murray River most of the way back to Melbourne. Many of the Aussie wives thought them mad, but they had a great experience and they enjoyed being the 'roughy toughies', comparing their journey to those of the men who had smoozed their way around the country in aircraft and air-conditioned comfort. Sue was able to make the trip through the kindness of the wife of my directing staff sponsor, Sally Burgess, who happily looked after Nick and Tim while Sue was away – yet another example of Australian generosity.

During the course of 1978, back in the UK, the Callaghan government had reacted to pressure to increase the pay of the Armed Forces and we received a significant, long overdue and welcome pay rise. In the following year Margaret Thatcher came to power and another weighty increase was added. In the course of twelve months, Armed Forces' pay went up by thirty-two per cent and from then on, although we were never particularly affluent, Sue and I were financially secure.

Australia was an exciting place to live and a land of extremes in many ways. During a particularly hot spell, a wind blew out of the centre, bringing with it millions, if not billions, of large, brown moths. Our houses all had fly screens on their doors and windows, but they were not moth-proof. Within minutes of their arrival the moths were everywhere inside and outside the house. If you opened a drawer, dozens would flutter out and for a day or two, when outside, you walked through something akin to a cloud of locusts. Just as suddenly, they were gone, but evidence of their presence was found on the beach. A black line, about a meter wide, stretched as far as the eye could see along the high water mark. The number of insects was incalculable.

The beach too was very different from those in Britain. There was nothing between us and Antarctica. Even in summer, cold squalls would blow in from the south, lowering the temperature by 10–15 degrees centigrade for a matter of minutes, until the squall had passed. It meant that you always kept a pullover handy. And the breakers were nothing like those in Cornwall. During stormy weather they were huge, magnificent and vastly powerful. The family loved the Point Lonsdale beach and its lighthouse with its booming foghorn, only five minutes' walk from our home, through tea tree, tussock grass and sand dunes.

After one storm, the four of us went to walk along the beach. We were careful to stay about 40–50 metres from the breaking rollers, walking next to a wall of sand that the sea had carved in the dunes. From nowhere a rogue wave broke and surged up the beach, sweeping all of us off our feet. Water boiled against the sand wall and an immensely powerful back tow carried us towards the sea. Sue grabbed Nick and, fortunately, the water released them both, but little Tim was swept past me. I caught him and we were both carried perhaps 10–20 metres towards the sea before the water gave us up. We were shaken and soaked, but otherwise none the worse having survived a truly terrifying incident.

The am-dram group was more active than ever. They had planned a much more elaborate production to be staged in the Staff College's lecture hall. However, the play they had selected would not be sufficiently long to occupy a full evening, so there was some pressure on the barbers' shop members to augment the drama production. Amongst the wives were two professional singers and a music teacher. They got together and came up with an edited version of *Joseph and the Amazing Technicolour Dreamcoat*. Even edited, that was a huge advance on the few songs sung earlier in the year. Auditions were held and I was cast as one of the brothers. Scenery makers, lighting and sound people were recruited for both the drama and musical productions. We needed musicians and eventually most of the Staff College and some of their family members were involved. We rehearsed twice a week and the success of a rehearsal was judged by the number of wine boxes consumed – four was considered acceptable.

The musical encountered a problem. The leading man was not quite up to singing his very large part and the replacement was similarly unsuccessful. I found myself in the limelight. The ladies in charge of the production would not take 'No' for an answer and I became Joseph by default. A programme of private tuition was arranged for me with the music teacher, lasting about two months. She worked me hard. As a young boy, in Wales, I had singing lessons and had sung in *eisteddfodau*, but had not sung since leaving school. Some of my early instruction came back to me and with the help of my mentor, my voice improved considerably.

The productions were put on over two nights to a packed hall. My mother had arrived from England in time for the second night's performance. She had been a choral singer virtually all her life and would be my harshest critic. I sang for her. She was delighted and I was relieved. '*I closed my eyes, drew back the curtain, to see for certain what I thought I knew*'. I can still remember the words.

During the second term's study break we had driven to stay for a few days in Canberra. My mother's visit coincided with the last of the study breaks and we decided on a final tour along the Great Ocean Road and Princes Highway to Adelaide, staying at well-appointed caravan parks on the way. For the return journey we took a northern route that followed the Murray River until we headed south to Bendigo and Ballarat to get back to Point Lonsdale. It was the first time that my mother had left England (she never did things by halves), and she was entirely unphased by the distances that she travelled. She simply enjoyed her family, a new country and its people.

For several months all students had been working on their Commandant's Paper. Because of the terrorist incident in Sydney at the beginning of the year and my experience in Northern Ireland, I selected *Para Military Forces* and their contribution to peacekeeping as my topic. Each student was given a directing staff mentor. Mine was Wing Commander Stan Clark, RAAF. We only had one meeting that lasted less than a minute. 'I don't know anything about para military forces,' was his only contribution, other than to buy me a drink and wish me good luck.

The course was reaching its climax. The final exercise was a TEWT – Tactical Exercise Without Troops – played out over about two days. As the name indicates, in this type of exercise no combat forces are deployed, but headquarters are, along with all the communications systems that support them. An exercise scenario was written and the player headquarters had to react to the situations presented to them. It was a Corps level exercise, set somewhere on the northern coast of Australia i.e. a lot of the ground would be 'Gafa', road communication would be elementary and long, making logistic support extremely difficult were the scenario to be played for real. Who the invading forces were, I have no idea. Throughout the year I could not see that Australia had a viable enemy – at that time only India possessed an aircraft carrier and no other south east Pacific nation had the means to deliver and support an invading force.

At the beginning of the TEWT I was appointed to be a divisional commander and an Australian mate of mine, Chris Roberts, was the Corps commander. He gave his orders. I gave mine and the exercise commenced. All went well for most of the first day, then I got an enraged telephone call from Chris asking why, in the name of all that's holy, had I not complied with his latest order? I said I had not received an order and that seemed to upset him even more. Relationships were strained for several hours, but whatever it was, we got over it; won our war; and have remained firm friends ever since.

Tony Redwood-Davies and I enjoyed one last moment of pure joy when the MCC side thoroughly defeated the Australians in the last test of the 1978 series. An entrepreneur, Kerry Packer, had 'stolen' most of Australia's better cricketers to play in a 'World Series' that he had organised. The remainder were no match for the MCC and they were thrashed five tests to one. A lot of students had assembled in the College's television room to watch the final rights being played out. The MCC won by an innings, or ten wickets, or something like that and Tony looked down the front row of seats at me and nodded towards the bar. I nodded my agreement and we both stood up to walk towards the room's exit. It was all too much for one of the Aussies whose frustration overcame him.

'Say something you bastards!'

Still no words from us and our broad smiles did not help him much either.

My final interview at the College was with the Commandant, Brigadier Kirkland. I did not know him at all well, but I thought him to be a blunt, joyless individual. He had taken over part way through the year, but somehow or other, the directing staff and the students had combined to negate his 'noses to the grindstone' message. Despite his efforts, 1978 had been an outstanding year for the Australian Staff College.

The Commandant gave me my report. It was a good one. I had got a 'B' grade and the right to have 'psc(a)' after my name – passed staff college (Australia). His criticism of me was that I lacked diplomacy. Talk about the kettle calling the pot black! Without another word he shook my hand and that was the professional side of the Staff College well and truly over.

All that was left was the final ball, held in the mess. It was a great party. Most of us saw the dawn come up from the old gun emplacements and, fittingly, we fired a volley of champagne corks towards the mouth of Port Phillip Bay. It was a happy, but bitter sweet event. We had had a great time with some wonderful people, but the following day would see the beginning of the exodus and many 'goodbyes' were painful. We packed our suitcases – the MFO boxes were already on their way to Berlin – cleaned the bungalow and flew to Sydney to catch a flight to Hong Kong. Nick cried bitterly as Sydney Airport receded below us. Now aged five, he was able to appreciate the value of friendships and he knew that he was unlikely to see his Australian playmates again.

In Hong Kong we had a couple of days sightseeing, before catching an RAF trooping flight back to the UK. The RAF looked after us well on a long flight that had short stopovers at Colombo, Sri Lanka and in Bahrain. We were about to land at RAF Brize Norton at about 8 o'clock on a chillingly damp evening, only to learn that the airfield was shrouded in thick fog. Conditions were marginal, but the pilot was going to attempt a landing and we strapped in tightly. I looked out of my window and saw nothing but darkness and the flashing lights from the aircraft's under belly reflecting on billowing fog. Everyone could sense the care being taken with the approach when, suddenly, the engines roared and the aircraft climbed steeply. A minute or so later the pilot's voice came over the speaker system.

'Sorry,' he said, 'I couldn't see the runway at all. I'm going to give it one more go and if that fails we will have to divert. The nearest airfield open is at Prestwick.'

Scotland!

Around we went to make a second approach. Out of my window I could see no break in the fog, but suddenly the black and white stripes at the end of the runway flashed past beneath us, almost within touching distance. The plane landed to an outbreak of relieved cheering. I signed the paperwork to secure a new car that was waiting for us. We loaded it up and drove to Sue's parents' home to spend a second successive Christmas with them. It hardly seemed possible that only a year had passed since we were there last.

On considered reflection, 1978 was, without any doubt, the most memorable twelve months of any that I and Sue spent with the Army. A huge 'thank you' to the staff, officers, wives and children that were the Class of '78.

BERLIN – A SEC

TRUE TO FORM, Sue and I were star
just like that to Londonderry, we d
to be the one to stay behind, while
first. The joys of a winter course a~ ~~~
awaited me.

In Berlin, I was to command 'C' Company, a rifle company of nearly 120 soldiers, and I had to attend a company commanders' course that would run for about three weeks. Such timely preparation was absolutely typical of the way the Army trains its people. I joined the Service in 1962 and was to leave it in 1996. It was not until 1997, after I had secured civilian employment, that I was not given training appropriate to my job. When people remark how efficient all three Armed Services are, they need look no further than the training investment that they make.

I arrived at the Students' Mess at Warminster on a cold, snowy day in the first week of January, driving our shiny, new Toyota Carina, leaving Sue to prepare for the air trooping flight to Berlin. Later, Sue wrote that she and the boys had made the journey successfully and were safely ensconced in a comfortable quarter. Her principal concern was that Berlin was having an exceptionally severe winter and cold weather clothing for her and the boys was difficult to locate – British stuff was simply not up to the job.

I remember little of the company commanders' course, except that the weather was the most testing element. At its conclusion the students dispersed on a Friday afternoon and I was looking forward to driving my new car across France, Belgium, Holland and West and East Germany, to be reunited with my family. I got as far as Reading.

I was approaching a roundabout when there was an almighty bang from the engine compartment and every dashboard light flashed and immediately died. My instant reaction was to brake and that brought a squeal of brakes and a booming blast on a horn from behind me. I looked in my rear view mirror and huge letters spelling VOLVO completely filled it. The Volvo truck's angry

my door to swear at me, but when I explained what had
ndly helped me push what I now regarded as a pile of Japanese
de of the road.

pposite the entrance to a manufacturing plant where a member of the
of Commissionaires was controlling traffic into and out of the factory.
ked him if I could use the phone in the gatehouse to ring for recovery. He
flatly refused. I looked at the medal ribbons on the left breast of his tunic.

I said, 'I bet you got that LS&GC [Long Service and Good Conduct medal]
for years of undetected crime. You probably owe a couple of good turns. Come
on, help me out.'

He gave me a sharp glance. 'Are you a soldier?'

Problem solved, moreover he told me that there was a very large Toyota
dealership only a mile or so away and came up with its telephone number.
A recovery truck arrived in about ten minutes and my car was delivered to
the most immaculate workshop I had ever seen. The place was spotless and
all the mechanics were dressed in white overalls – also spotless. I was asked
to describe what had happened and handed the car keys to the mechanic
allocated to investigate the problem. He spent barely a minute under the
bonnet before shouting for his mates. For the next thirty minutes three or
four mechanics swarmed over the engine compartment, dismantling the
engine almost completely.

The dealership's manager was called to be briefed. He approached me to
say that a tappet had fallen into the engine bloc and that the engine was a
write-off. I explained that I was on my way to Berlin and had a ferry to catch
later that evening. He sympathised and said that he would make the building
of an engine for my car his absolute priority, but because Toyota held no
major assemblies, a replacement engine would have to be built from scratch. It
would take all weekend to assemble the parts from Toyota outlets around the
country and the engine would be built on Monday, in time for me to catch
an evening ferry from Dover. I was given a very large courtesy car and drove
to my mother's home in Sudbury to spend an unexpected weekend with her.

On Monday afternoon, I returned to Toyota's 'operating theatre' in Reading.
My Toyota was ready for me, full of petrol and gleaming inside and out. The
manager was profusely apologetic and told me that my failed engine bloc had
been flown to Japan over the weekend. I was impressed with the service and

An early family photograph taken immediately before leaving for Londonderry. December 1975.

Sue (holding another officer's hand), after her helicopter ride to our farewell party. Magilligan Point, County Londonderry. September 1977.

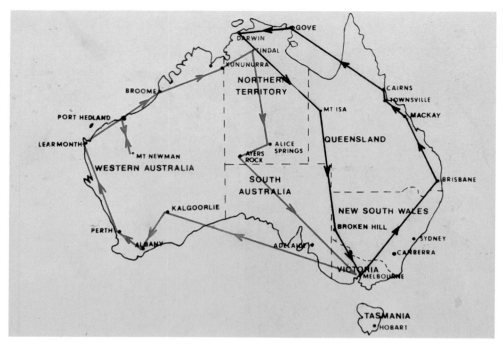

Map showing the routes of the Australian Staff College's strategic tours of 1978.

Photograph of the Australian interior taken from the open back ramp of a Royal Australian Air Force C-130 (Hercules) plane.

Me, 'strutting the boards' as 'Joseph'. Australian Staff College production 1978.

Frontispiece Allied Forces Day Parade programme. Berlin 1979.

The Queen's Birthday parade held on the Maifeld (1936 Olympic equestrian venue) with the Olympic stadium in the background. Berlin 1979.

Dress rehearsal of a re-enactment of the 'Battle of Sobraon'. The Poachers soldiers get stuck in at Montgomery Barracks. The sports fields were only 200m from the border with the DDR. Berlin 1979.

Bobby Roberts with Mary (lead elephant) about to rehearse the re-enactment of the Battle of Sobraon at Montgomery Barracks. Berlin 1979.

Berlin Travel Document for me and my family with Soviet Army stamps that permitted exit and entry from and to Berlin and West Germany. July-August 1979.

'Trooping the Colour'. Leading the Pompadours parading to celebrate the Regiment's tercentenary. Ensign – Lieutenant David Moorat. Adjutant – Captain David Clements. 17th August 1985.

Explaining the finer points of a Challenger tank to the Rt Hon George Younger, Secretary of State for Defence, during his visit to the Pompadours' battlegroup. Soltau, West Germany 1986.

The Pompadours battlegroup drawn up on the Alberta prairie, prior to undergoing weeks of extensive live fire training. The reconnaissance elements are in the foreground; two squadrons of tanks ring their associated infantry companies in the large box of fighting vehicles, with battlegroup headquarters, the Anti-tank and Mortar Platoons in the centre. The rear rank of the main box is formed by Engineer assets, including two bridge layer tanks which dominate. Behind them are the vital logistic support elements. May 1986.

My sons, sister-in-law and me about to enter the DDR, prior to unification, along a hastily laid track, through a gap cut in the border fence. July 1990.

Formal photograph taken on my first day as a brigadier. January 1991.

Visiting troops in Northern Ireland. 1992

Newmarket House. The Commandant's residence at Sennelager. A significant upgrade from our first quarter in Warminster. January 1991 – July 1993.

MINISTRY OF DEFENCE
MAIN BUILDING WHITEHALL LONDON SW1A 2HB

Telephone 0171-21......................(Direct Dialling)
0171-21 89000 (Switchboard)

PARLIAMENTARY UNDER-SECRETARY OF STATE
FOR DEFENCE

30 April 1996

Dear Brigadier Groves,

The Secretary of State for Defence has it in command from Her Majesty The Queen to convey to you on leaving the Active List of the Army Her thanks for your long and valuable Service.

May I take this opportunity of wishing you all good fortune in the future.

Yours sincerely,

Howe

THE EARL HOWE

Brigadier C Groves

Recycled Paper

Farewell letter from the Parliamentary Under-Secretary of State for Defence on leaving the Army. 30 April 1996.

the concern shown. We kept our Carina for several years after that incident with not a hint of a further problem. There was a sequel – several months after my breakdown in Reading, I was sent a copy of Toyota's in-house magazine and on the front cover was a picture of my faulty engine with the company's investigation related on its inside pages. Obviously, Toyota took its reputation extremely seriously.

My journey to Berlin was long and icy. The procedure to get past the Russians, going into and leaving the DDR, had not changed one iota and I felt at home entering the western half of the city. My new unit was the 2nd Battalion, The Royal Anglian Regiment (the Poachers) and they were in Montgomery Barracks, in Kladow, at the extreme south west of the British sector. In fact, it was impossible to go any further in the sector because the border with the DDR formed the perimeter of the barracks, except in one small area, where one derelict barrack block was actually in the DDR. At that point the East Germans had set the border wall back about 150m. The border also ran just beyond the end of the street in which our quarter was situated. As a consequence, Nick and Tim had one of the most secure play areas anywhere in the world.

Changing battalions was a big deal for me. I now wore a black lanyard instead of my Pompadour purple one and my Eagle and Garter collar badges were put away in favour of the 2nd Battalion's Sphinx. On my first day with the battalion I made my way to battalion headquarters and the adjutant showed me into the commanding officer's office. I did not know Lieutenant Colonel Patrick Stone. He had served his early career with the 1st Battalion and our paths had never crossed. I marched smartly into his office; stamped to a halt; saluted and said, 'Major Groves, reporting for duty, sir.' He was writing as I entered: he continued to write and did not look up. As a company commander I was now amongst the most senior people in the battalion and I thought that his overly cool reception represented a marked lack of manners. The CO completed his writing, put his pen down and finally looked at me. I expected a, 'Welcome, Colin' or 'How nice to see you', but not a bit of it.

His first words were 'What do you know about elephants?'

There was a prolonged pause. 'OK,' I thought, 'I'll play his game.'

'They're big and they're grey, sir,' I said.

'You're the expert I've been looking for' – his response came far too quickly. He had set me up!

Patrick Stone was a gentleman and it was a pleasure to serve under him. He was a caring soldier, with a wry sense of humour. He welcomed me warmly and went on to explain that he had drawn the short straw when it came to contributing to the British Military Tattoo in Berlin (akin to the Royal Tournament, but a bigger production). He had been told to produce a re-enactment of the Battle of Sobraon – a battle fought in 1846 as part of the Anglo-Sikh wars, where a forerunner regiment of the 2nd Battalion distinguished itself. He told me to use my imagination and devise an extravaganza; that I would have most of the battalion at my disposal and that he wanted cannons, horses, camels and elephants!

'What we must do,' he explained, 'is make our idea a total impossibility and then the organisers will forget us and tell the Grenadier Guards to march up and down a bit. We'll be off the hook!'

The Staff College course and the last weeks running around freezing on Salisbury Plain had done nothing to prepare me for this and I wondered what I had fallen into.

I left battalion headquarters to find 'C' Company's accommodation. As I walked through the barracks I was immediately impressed. For my previous tour in Berlin, the Pompadours had been accommodated in a 'Kaiser' barracks, situated in a residential area in Spandau and it showed the wear and tear of the better part of a century. By contrast, Montgomery Barracks had been built by Goering, for Luftwaffe troops, between the two World Wars. Goering had not skimped when constructing it and its buildings looked as pristine as they would have done in the 1930s. The barracks was set in a big, open area and was screened from local residential housing and the border with the DDR, by attractive pine woodland. Apart from low, two-storey accommodation blocks, messes, offices, garages and an ammunition compound, it also contained an indoor range, two outdoor ranges (30-metre and 200-metre), an assault course, gymnasium, swimming pool, football, rugby and hockey pitches and a running track, all overlooked by an excellent sports pavilion.

I did not know anyone in my new company and was denied a direct takeover from my predecessor because he had already left to take up his next appointment. The company second-in-command, Captain Kevin Hodgson, was the interregnum. He introduced the main personalities to me and showed me what had been planned for the next couple of weeks. It consisted of individual, section and platoon training, followed by a round of guard duties. It was exactly what I had experienced fifteen years previously. From a military point of view, Berlin had not changed.

Straightaway it was apparent that I had inherited an efficient, well trained company, but I gave myself two weeks for unobtrusive observation, to find out exactly what the officers, NCOs and men were made of before attempting to put my stamp on the outfit. That fortnight also gave me the chance to work on the Sobraon project.

The Tattoo was to be held in the Deutschlandhalle, Berlin's equivalent of London's Olympia, so I had a big arena to work with. I researched the battle to get some idea of its ebb and flow, the uniforms, weaponry, topography and, hopefully, the use of animals. I could not find any references to camels and elephants, but who was I to let truth stand in the way of a good story? I decided that the Redcoats would storm ramparts and gates; they would have a cannon to assist them; ensigns would carry the old six foot square Colours; and British attacks would be repulsed a couple of times before they made an heroic breakthrough through the Sikh defences. The British commanders would be mounted and the elephants would be employed to pull the cannon, before helping to break down the gates – the stuff of Hollywood. Camels were abandoned as a step too far. Patrick Stone was delighted. He was certain that the powers that be would be entirely put off by my ridiculous concept.

I was instructed to work up a presentation for delivery to Michael Parker – who has since masterminded most of the UK's big national celebrations – and to the Tattoo's principal staff officer. The presentation was made in the commanding officer's office. While I was speaking I saw a look of pure incredulity spread over the staff officer's face. Michael Parker gave nothing away. I brought the presentation to an end and there was a moment of silence, broken only when Michael Parker unfolded from his chair and walked to the commanding officer, who was sitting at his desk. Michael extended his hand to Patrick Stone.

'At last, someone who is thinking at the right level. Brilliant. Go for it!'

He turned and left with the poor staff officer saying, 'Where are we going to get all the kit they'll need, let alone the bloody elephants?'

Patrick Stone looked at me and said, 'You wrote it – now get on and do it.'

Great! First of all I needed help with the production, so I looked to find two officers to command the Redcoats from horseback. They would become my assistants. Only one officer – Mike Beard – admitted to ever sitting on a horse, most claimed an equine allergy. I had had half a dozen riding lessons in Australia, so it looked as if I had written myself into a starring role.

The Tattoo was still months away and there were more immediate things that demanded my attention. I was keen to crank up the company's training programme a couple of notches. The soldiers were reasonably fit, but could have been fitter. All the basic skills were being covered, but the level of instruction from some of the junior NCOs required improvement and there was a need for a few high points in the programme; things for soldiers to look forward to doing. Fitness was easy to address and because I had some very good NCO instructors, I made them mentors for the weaker ones. I made the most of a village complex that the British had constructed at Ruhleben, to practise fighting in built up areas; more watermanship training was included because Berlin has vast stretches of water that would interfere with any military operation; orienteering was preferred to simple map reading exercises; and occasionally, and totally unexpectedly, realistic 'casualties' would require first aid. With those and other tweaks, I aimed to keep my boys on their toes and I had plenty of time to plan all that because when it was my company's turn to provide the guarding commitment, it left only a handful of soldiers, who did not require much of my attention to keep them gainfully employed. If I am to be honest, a company commander's life in Berlin was a bit of a 'cushy' number for some of the time.

I made the most of that free time to get back into serious sport. I became the battalion goalkeeper once again and when summer came around, I resurrected the battalion's athletics squad. I also rode, not because I particularly wanted to, but because now I was committed to riding in public, in a mock battle, with bangs and smoke and wild animals all combining to make any horse misbehave. I quickly realised that the only horses equipped for this job were West Berlin police horses. The police authorities were approached and like everyone else, they thought we were mad, but they were gracious in humouring the Brits. I reported to the police riding school in Spandau and stepped back in time. The riding master had waxed moustaches and was dressed in a frock coat, with an Uhlan cavalryman's box hat. He watched me ride and then folded my stirrups over the saddle to make me use my legs properly. I was never to get those stirrups back. My bum and thighs were sore for weeks, but I was pretty good at dressage by the time he had finished with me.

At home Sue was in some sort of seventh heaven. Our semi-detached quarter was comfortable and well maintained, with three bedrooms, a large cellar, garage and generous rear garden. I qualified for a daily cleaner, who also was available for babysitting. We had most of our groceries delivered because the British operated a system to turn over the huge supply of food kept in Berlin against the possibility of another blockade. All our families benefitted from it simply by filling in a form that was left in a basket at the front door. Baskets and forms were collected three times a week, when deliveries were made from refrigerated vehicles and blank forms would be left for the families' next orders.

Nick was at a forward thinking primary school, run by the British Forces Education Service, at RAF Gatow. The school building was very modern and a 'home base' system was employed instead of traditional classrooms. It was advanced for its time and the attraction of spending a couple of years in Berlin meant that good, young, British teachers were employed there. Tim was at the battalion's kindergarten in Montgomery Barracks, interestingly, located in the last building in the British Sector. When the tiny tots did their fire drill, they exited through a door into East Germany, only to run around the building back into the West.

For the first time for several years, Sue had some time to herself, but she really would have preferred a job. When one was not forthcoming, she took matters into her own hands and started a franchise business selling Soda Stream machines and the fizzy drink concentrates that went with them. Her enterprise took off, which gave her considerable satisfaction and a modest income.

That was not the only source of new income for us. A package arrived from Australia, from the publishers of the *Pacific Defence Review* magazine. Their letter contained the good news that my paper on *Para Military Forces* had been selected for publication and the relevant editions of the magazine were enclosed. Better still, the package also contained a cheque for several hundred Aussie dollars.

The Tattoo was now only weeks away. I had been put in contact with Bobby Robert's circus and Bobby had committed to bring his elephants to Berlin. Cannon, ramparts and gates had been designed and were under construction in the REME's workshops. Uniforms for the British and Sikh contingents had been leased from a company that normally supplied the film industry; I had liaised with a re-enactment regiment in America about ancient drill; and the armourers modified our rifles to look like muskets.

A forerunner regiment of the 2nd Battalion, the 10th of Foot, had fired the first shot of the American War of Independence, on Lexington Green. In the United States in 1976, to celebrate the bi-centennial of independence, there had been a clamour for re-enactments of eighteenth century battles and a group of Boston businessmen had decided that they wanted to play the Brits. Subsequently, they contacted our Regimental Headquarters for information and advice. They quickly became His Majesty's 10th Regiment of Foot of America and formed a close alliance with The Royal Anglian Regiment. By 1979 they had become experts on the drill of the time and a few of them paid their way to Berlin to help us produce some authentic musket drill for our battle in the Deutschlandhalle.

The elephants arrived in Berlin having travelled through Soviet checkpoints, complete with formal travel documents. They were accommodated in garages and were given Army numbers so that we were properly authorised to provide them with feed. They ate a lot, which meant that a corresponding amount came out of the other end. Once our rose gardens had reached saturation point, the residue was heaved over the Wall.

For the first rehearsal, soldiers were gathered on a sports pitch, together with animals, cannon and scaffolding representing ramparts and gates. The horses had been ridden out beforehand to tire them, but they hated the elephants on first sight. Even with a fistful of curbs and snaffles, it was difficult to control them and they bounced or circled around excitedly. The soldiers immediately named my horse Zebedee and Mike Beard's became Dougal – obviously the boys were of The Magic Roundabout generation.

Early rehearsals went catastrophically badly. We were using two elephants to pull the cannon, which was not a natural thing for them to do. Despite a lot of pre-training, almost immediately after being hitched to the cannons for the first time, the lead elephant bolted across the sports pitch scattering soldiers, to race up a slope into pinewoods and crash into the border fence with the DDR! Fortunately the cannon became detached in the process and was undamaged and Mary, the lead elephant, was easily recovered. During the next rehearsal the horses spooked and it was their turn to take off, down a steep bank, across the athletics track and football field and neither Mike nor I could make the brakes work. We solved the problem by riding into the ammunition compound that had a high fence. The gates were closed behind us and eventually the horses quietened and we could dismount.

Things improved and after a week of intense rehearsal, we had a presentable act. I was concerned that it would be a problem to get the elephants from our Kladow barracks to the Deutschlandhalle – 10 miles away – efficiently enough to comply with the movement schedule that had been written for us. West Berlin police motorcycle outriders were to escort our convoy and the whole thing was timed to the minute. I asked Bobby Roberts if moving elephants posed a problem. He laughed and had a handler place one of the drums that elephants sometimes stand on to preform tricks, next to his specially built elephant trailer. Bobby shouted 'Mary – get in the bloody truck'. Mary lifted her head and, as good as gold, she led her troop into the trailer. They all knew their places and each lifted a foreleg to have a chain attached for transit. Everyone, who had anything to do with the elephants, was incredibly impressed with their intelligence and their touching need for attention. They were truly remarkable animals.

The Tattoo was a sell-out. After a promotional parade along the Kurfurstendamm, West Berlin's main shopping street, and a photo opportunity with the elephants crossing the Havel lake on a Royal Engineers' heavy ferry, tickets sold like hot cakes. The Tattoo ran for a week with matinee and evening performances. Its varied programme included motorcycle, Scottish dancing and gymnastic displays, but the three stars of the show were undoubtedly the massed bands, totalling just over 1,000 musicians (their music reverberated through the building to surround and press in on you); the immaculate drill of the RAF's Queen's Colour Squadron; and … us!

The Germans loved the re-enacted battle. They cheered and booed as if at a British pantomime – perhaps they were! If we had reached a presentable level at rehearsals, the boys raised their game for performances. The still skittish horses sashayed into the arena – Mike and I dismounted as soon as possible to fight on foot – much, much safer; the cannons trundled on behind their elephants to boom away at the ramparts; the Redcoats charged and charged again; finally, the defences were breached and the Union flag was raised in an epic finale. The best bit was the close quarter battle between individual Redcoats and Sikhs. Initially, I had tried to choreograph it, but my attempts lacked lustre. My sergeant major suggested that we pair up groups of about four or five Brits with a similar number of Sikhs and let them sort out their own mini wars. The result appeared to be uncontrolled mayhem and a great spectacle. When it comes to punch ups, Tommy Atkins knows no equal.

The Tattoo's funniest moment came on its final night. The Queen's Colour Squadron were professional tattoo performers. They were exceptionally good and knew it. The climax of their performance came when all the house and arena lights went out and they marched in the dark, with only their luminescent gloves and rifle slings indicating their perfectly synchronised drill. The lights had to come back on the instant they fired a volley. They fired as one; the arena was illuminated and out of sight, somewhere in the rafters, the Royal Signals lighting technicians dropped two dead ducks. Perfect! The audience thought it was part of the show and applauded wildly. The RAF were poh-faced.

The annual programme for British infantry battalions posted to West Berlin was very structured and saw them move from one major event to another. The first months of the year were taken up with training for the platoon battle tests; joint exercises were held with the French and Americans; there were always two major training periods in West Germany – one for live firing, the other for tactical manoeuvre; and the early summer months were given over to drill for the Allied Forces Day parade – held along an impressively wide boulevard, leading from the Brandenburg Gate to the Victory Column on Strasse des 17 Juni, and for the Queen's Birthday Parade – still held in the Olympic equestrian arena. That programming made for easy personal planning. Over two years, my family was able to take summer holidays, first in the Dordogne – where Nick and Tim developed an early taste for fine dining that has never left them, and then to Denmark, where our holiday accommodation was unusual – on a pig farm, in the north of the country, near Alburg. We made visits to a wide variety of attractions from Legoland, to the bog man of Aarhus and to the sea, where the Skagerrak meets the Kattegat. Often people would claim that being cooped up in West Berlin made them feel claustrophobic. I never suffered that, but certainly I felt greater freedom during times spent out of the city and our holidays were all the more enjoyable because of it.

My second year with 'C' Company started well for us all. The platoon battle test competition was on us again and I relished the opportunity of having an intense period of training to organise. I put my lads through the mill and they responded wonderfully to every tactical and physical challenge that I set. Things were made even tougher for them because yet another very cold winter saw night time temperatures fall below minus 20 degrees centigrade throughout the preparation training and for the event. Twenty-seven platoons took part and I was confident that my three platoons would perform well. What I did not expect was that they would come 1st, 2nd and 7th and the platoon that came 7th would have come 3rd had not a soldier become a casualty on the last day, incurring penalty points that pushed his platoon down the pecking order. It was a wonderful result that was celebrated commensurately.

If Londonderry had been something of a cultural desert, then Sue and I were utterly spoilt for choice in Berlin. The range of museums and galleries in the city match those of London. Additionally, we made the most of the opera, concerts and theatre. If we went to the opera it was normally in a group, to the Staatsoper in East Berlin, situated on the Unter den Linden. The DDR's government poured enormous amounts of money into the arts and, as a consequence, productions in the East were on a grand scale. Invariably, after an opera, we would go on to eat at the Ganymed restaurant (also in the East) and that was an entertainment in its own right. It had reasonable food; unknown wines of varying quality from eastern bloc countries; a geriatric string trio; and, word was, that it had Stasi bugs all over the place. We had great fun talking absolute rubbish in the hope that some unfortunate Stasi operative would have to spend long hours trying to work out the importance of what we had said. In the West, we attended outstanding concerts in the Philharmonie to hear such stars as James Galway, Harry Belafonte and the Berlin Philharmonic, under Herbert von Karajan. But best of all was a production of the sinking of the Titanic in West Berlin's Opera House.

We arrived at the theatre only to find a crowd milling around outside. None of the theatre's doors were open. With only minutes before the production was due to start, men dressed as sailors approached the theatre pushing a gangplank. More sailors appeared above us, on the first floor of the opera house, to remove a huge glass window. The gangplank was positioned to give access through the window space and the audience was directed through that unexpected entrance. It soon dawned on us that the entire theatre was being used to represent the ship. The audience only got as far as the bar and the walkways outside the auditorium – they were served drinks and hors d'oeuvre there. Then an orchestra began to play and members of the cast, dressed as passengers, invited some of the real audience to dance. The rest of us 'cottoned on' quickly and soon everyone was dancing and enjoying that very original start to the production. Suddenly, there was a tremendous crash and the lights went out. We had hit the iceberg! Immediately, sailors with torches began organising us.

We were first class passengers and were to be evacuated before anyone else. We were led, in groups, through the deserted and darkened auditorium, up and over the stage, and then down into the bowels of the theatre, past steerage passengers (actors), screaming to be allowed out of their accommodation and above deck. At various points along the way, we would be stopped and there would be some music and singing to support the action. Finally, we were in the open air, but in a courtyard that was hemmed in by tall buildings, perhaps six to eight storeys high. We were still in our groups, which matched the numbers that would occupy a lifeboat. Darkness had fallen since we had made our entrance up the gangplank and to give an uncomfortably realistic representation of the motion of the sea, searchlights played up and down rhythmically on the buildings' facades. The music and singing continued until only a violin was playing. The searchlights moved slowly up the buildings, leaving just a funnel illuminated on the roof of one building. The violin's music stopped in the middle of a piece and the light on the funnel went out. The Titanic had sunk. There was a moment of absolute quiet. The imaginative representation had vividly recreated drama and loss – then the applause broke out and went on and on. It was an absolutely memorable performance.

West Berlin offered the allied powers a unique benefit with regard to tactical training. If war against Warsaw Pact forces were ever to come about, most of the fighting would be in built up areas, which is a particularly bloody form of conflict and devilishly difficult to control. The British had the fighting village at Ruhleben in which to practise, but its scale was minute compared to the residential and commercial areas that surrounded it. The village was sufficient to develop the basic skills of forcing an entry into a building and clearing its rooms of enemy, but it did not allow practise of protracted operations, over a wide area and with really big buildings to defend or fight through. In West Berlin, the solution was that when large buildings, such as apartment blocks, or factories, were scheduled for demolition, they were turned over to the military until the time came for them to be knocked down. I was to have use of two or three such buildings, including an underground station. As a result 'C' Company became quite proficient at FIBUA (fighting in built-up areas) operations, to the point that my company was selected to provide a demonstration to an audience of officers, not just from the allied powers in the city, but from the BAOR and from the United Kingdom. It was the centrepiece of a high profile FIBUA symposium and the pressure was on us to get it right.

Two weeks before the event, we took over a derelict factory complex and prepared it for defence. About ten days and 30,000 sandbags later we were ready. The sandbags protected fire positions in windows and reinforced floors. 'Mouse holes' were knocked through walls to allow defenders to move from room to room without being seen. Booby traps and mines were laid where attackers might well try to make an entry and smaller buildings were flattened to deny cover to an opposing force and to provide us with cleared fields of fire. We rehearsed our defence against an enemy drawn from other companies of the battalion until the demonstration was really polished. The whole thing was on a scale that none of us had seen before and it served to underline how difficult and horrific FIBUA operations would be. While we were pleased to have been involved in such a successful event, the probable level of casualties was a salutatory lesson for everyone.

Military units are very tight knit – the smaller the unit, the tighter the bonds. A rifle company has a little over one hundred personnel, so everybody knew everybody. My sergeant major came to see me one day to tell me that one of my soldiers was getting married in only three days' time to a young German lady. I cannot recall the name of the soldier, but I remember that he was one of the company's characters, often distinguishing himself in training, only to blow it all by getting into repeated scrapes in his free time. He had done absolutely nothing about getting married, other than to arrange a registry office ceremony. We just could not let that happen. The padre was alerted and a service was arranged in the battalion's chapel to take place after the civil ceremony. The bandmaster sorted out a quartet and some suitable music. The senior chef conjured up a buffet menu and I gave the company a day off to be free to celebrate with the bride and groom. A whip round paid for it all. It was a truly lovely occasion. The young German woman seemed not to have any close family, just a few friends. I was privileged to give her away with the company in attendance After the service, we 'borrowed' the commanding officer's staff car to drive them away, but they only made one circuit of the barracks before arriving at the company's club for the reception. That went on for some time and developed into a disco in the evening. It was a great day and a real family occasion.

1980 rolled on through the parade season and training periods in West Germany. It was an extremely good sporting year for the battalion, with the football team reaching the final of the BAOR leg of the Infantry Cup, and the athletics squad matching that effort to qualify for the BAOR inter-unit final. Commanding officers changed and Patrick Stone was succeeded by Roger Howe. Roger had been my predecessor as the commander of 'C' Company and so knew the battalion extremely well. Two important announcements were made – first, that the battalion would be posted to Northern Ireland early in 1981; and second, my next tour of duty would be at the MOD, beginning in December.

Roger Howe was determined to get his 'pound of flesh' out of me before I departed. He made me President of the Mess Committee (PMC) for my last few months serving with the battalion. The PMC ran the officers' mess as an 'add on' to his normal job, with the aid of a committee and the mess staff. I had avoided the appointment like the plague, so had no grounds for objection. There was a big bonus. Other than Eltham Palace, in south east London, where I had stayed briefly at the beginning of my career, the officers' mess in Montgomery Barracks was the most impressive that I had ever seen. It was stone built, largely oak panelled, with generously proportioned rooms and it was only about 20 metres from the DDR's border fence. If I put my mind to it, I could hold a memorable party there before I left.

I did put my mind to it and a medieval feast was staged. The chefs thoroughly enjoyed catering for something different. I needed suitable entertainment and the band rose to the occasion. So too did the physical training instructor, a specialist SNCO from the Army Physical Training Corps. He was a tumbler by profession and his display was outstanding. And there had to be miscreants (any subaltern officer would do) and stocks to hold them while they received their punishment. The battalion's domestic pioneers (carpenters and painters) sorted the stocks out in a morning. The mess waiters could not be left out and in the Berlin Tattoo's stores I located bottle green replica uniforms of the Royal Company of Archers – the Sovereign's bodyguard in Scotland – and if they were good enough for Her, they would be good enough for me. Finally, I had to have guests. The battalion's officers, their wives and girlfriends crammed in and, for outside guests, an invitation to the feast was the hottest ticket in Berlin. On the night, the guests' costumes were fantastic. I borrowed the assault pioneer sergeant's ceremonial axe and went as an executioner to discourage any complaints. It is very pleasing that the medieval feast has gone down in 2nd Battalion history.

My last days at duty with the Poachers were spent at Sennelager Training Centre in West Germany. Its replica Northern Ireland village that the Pompadours had used prior to their deployment to Londonderry in 1973, had developed exponentially. The village now had perhaps sixty to eighty buildings, representing houses, shops, garages and pubs. The village and other specialist training facilities were staffed by members of the Northern Ireland Training and Advisory Team, all of whom had recent experience in the Province. No longer did commanding officers have to devise their own training programmes; a constantly updated, tailored package was delivered to them.

The sophistication of the training was impressive and reassuring. Soldiers felt that at the end of it, they could not be better prepared for the task ahead. Companies rotated through live firing ranges, patrolling exercises, checkpoint, house and vehicle search techniques, first aid training and follow-up procedures for shootings and bombings. The village brought all of those things together over a three-day period. It was a real test. By the end of it I was confident that 'C' Company was ready and very much regretted that I would not be leading them in Northern Ireland.

Back in Berlin, my company saw me off in some style. I cannot remember much of that party, but I was privileged to have served with some first-rate officers and men. I remember the Poachers fondly.

West Berlin was a remarkable place to serve. It remained besieged and a beacon for western democracy. I had been unusually fortunate to have had a second chance to experience its unique atmosphere and to do my small bit to preserve its freedom.

I left the city to drive to England alone, with the car heavily loaded as usual. Sue and the boys would fly back to the UK a few days later. I was headed to take over a quarter in another 'WB', this time West Byfleet. Somehow it failed to have quite the same ring as West Berlin.

CHAPTER 13
SOLDIERING IN A SUIT

THE JOURNEY FROM West Berlin to Surrey was tediously long and I was relieved to reach the Army quarters in West Byfleet. Serendipity had intervened and I was to stay overnight with Nigel and Mary Still (Nigel had been one the UK's exchange directing staff at the Australian Staff College in 1978). They were to be our neighbours for the next couple of years. An even greater coincidence was that Compton Boyd – the other of the UK's exchange directing staff officers at Queenscliff – and his wife Biddy, now lived only two further doors away.

I looked out of Nigel's front room window, across the road to 12 Dodds Crescent, the quarter that Sue and I had been allocated for the next two years. It was a tiny, semi-detached house exactly like the one we had 'camped' in at Shrivenham some three years previously. I did not look forward to breaking the news to Sue. We were both going to miss our generous quarter in Berlin, especially when most of the other quarters around us were so much bigger and the civilian residential area that adjoined the Army patch, was extremely smart.

When my job at the MOD had been announced, Roger Howe, the CO of the 2nd Battalion, had been very enthusiastic. 'You've landed a really good one there,' was his gist. My job title was 'GSO2 ASD2d'. If readers have no clue what that could mean, I was very much in the same boat, except that I knew 'GSO2' meant General Staff Officer Grade 2 – a major's appointment. Roger began by explaining that I would work on the 5th Floor of the MOD's Main Building in Whitehall.

'What's good about the 5th Floor?' I wanted to know.

'It is directly below the offices of the Chief of the General Staff (CGS) [head soldier] and the other MOD-based Army Board members [all generals]. Military Operations and Army Staff Duties (ASD) are the principal staff branches that support them.'

What about the '2d' bit. Roger explained that ASD2 was the Army's implementation branch. When an Army operation had been approved at governmental level, ASD2 took over to allocate the forces required to carry it out – that sounded interesting work. The 'd' would have to wait until I got to London.

After breakfast with the Stills, I took over the quarter, unpacked the car, made up the beds with sheets and blankets from the 'get you in' pack – a range of essentials that the Army provided to enable families to negotiate the first few days until their belongings caught up with them – bought some flowers and made the place as attractive as possible to receive Sue and the boys who were flying to Luton and then would travel by rail to West Byfleet. I picked them up at the station that evening and we were quickly installed in our new home.

I was required at the MOD immediately in order that the takeover from my predecessor would be complete before the Christmas holiday. Out of uniform and in a suit, I caught a rickety, slam-door, commuter train from West Byfleet at about 07.40 in the morning. If all went well it would get me to Waterloo by 08.10ish and I could walk across Hungerford Bridge, then along Whitehall Place to the MOD and be at my new desk by 08.30. Needless to say, those timings were subject to British Rail and for the next couple of years anything above a fifty per cent success rate would be a bonus.

My predecessor was Henry Wilson, an Irish Guardsman. He was delighted to see me and was very much looking forward to returning to regimental duty – worryingly so. The 'd' desk turned out to be responsible for operational deployments in the UK. I cannot remember the order in which the responsibilities came, or the exact designations, but the 'a', 'b' and 'c' desks looked after NATO, Middle and Far East, and Rest of the World deployments. Each of the other desks had a GSO2 and a GSO3, but I had a second GSO3, who, with the IRA's campaign at its height, was entirely Northern Ireland orientated.

On the first day, I had a round of interviews with various bosses. My immediate boss was Colonel Tony Makepiece-Warne, a Light Infantry soldier with a waspish sense of humour. I had landed on my feet with him. Next came the Director of Army Staff Duties, a major general, a 'two star', whose name escapes me. He and the Director of Military Operations were the two most powerful major generals in the MOD – one oversaw the design of operations; the other implemented them. After his welcome and overview, I began to grasp the importance of serving in ASD2. Finally, I was not introduced, but re-acquainted with the Deputy Director of Army Staff Duties, a brigadier and none other than Keith Burch, the most unpopular commanding officer the

Pompadours had ever endured! However, there was an immediate surprise in store for me – away from regimental soldiering and in a staff environment, Keith Burch was brilliant. He had an immediate grasp of the complex issues that confronted us and he readily offered support and advice to all his subordinates. His was a chameleon-like change and one that I could never have predicted.

In the days that followed, Henry Wilson went through a very thorough and diverse handover. Virtually everything ASD2 was involved with was classified, more often than not, extremely highly classified. He introduced me to MACA that was one of the main aspects of the job. The acronym stands for Military Assistance to Civil Authorities and that divides into three distinct areas.

Military Aid to the Civil Power (MACP) – essentially to reinforce the police in cases of extreme civil disobedience, or to provide resources they do not have e.g. if terrorism can be regarded as civil disobedience, then Northern Ireland was a case in point, but a lesser MACP request might mean deployment of a bomb disposal team to deal with a World War II explosive device.

Military Aid to Other [Civil] Ministries/Departments (MACM) – usually meant deployments to help alleviate the effect of industrial strikes that threaten the health, safety and wellbeing of the community at large. Cover for striking fire and ambulance services are the best-known examples. Current MACM operations are the military involvement in the building of Nightingale hospitals, the provision of Covid-19 testing teams and the distribution and inoculation of vaccines.

Military Aid to the Civil Community (MACC) – most commonly deployments to help with natural disasters like flooding.

During the latter stages of the handover week, Henry and I went to Northern Ireland where I was briefed in detail on the current security situation at Headquarters Northern Ireland and on possible developments. I was also briefed on the operations of the Special Patrol Group (SPG), which conducted covert operations in the Province. I was the staff officer who represented their interests at MOD and would later make a two-day visit to one of their operational detachments.

Once back in England, Henry booked an appointment with officials at the Cabinet Office, but would not tell me why I would be involved with that department. On the Friday evening we went for a drink and then shook hands – I now had the job. In that instant and amongst the other responsibilities that

I had assumed, I became the legal governor of Durham Jail. There was a prison officers' strike in progress. Durham Jail had yet to be commissioned and the military had taken it over to alleviate overcrowding in other prisons. As the author of that plan the GSO2 ASD2d became the legal governor. Additionally, the plans for assistance in ambulance and fire service strikes had been dusted off; industrial relations with the rail unions were acrimonious; the water workers were unhappy; and the security situation in Northern Ireland continued as ghastly as ever. No wonder Henry was happy to return to his battalion.

If I thought my first weekends with my family in West Byfleet would be restful, I had another think coming. Right at the top of Sue's to do list was plumbing in our washing machine. No plumber was available; the task fell to me and I am not known for my DIY skills. I bought the fitments, found the water main, turned it off and started work. All did not go well; my temperature rose as this seemingly simple job took several hours. At last it was done and I could relax. Not a bit of it. Over the next weekends the bathroom flooring needed attention; our furniture had to be moved in from store; the corresponding unwanted items from the Army had to be returned; and the contents of the MFO boxes, temporarily stored in the garage, had to be unpacked. When those boxes finally revealed their contents, it was evident that only about half would fit in the living areas of the house. That was where the attic came into play and I was to become closely acquainted with it, when, several times in a year, I would balance on a step ladder to remove and replace seasonal items from the attic, for example, swapping winter duvets for garden chairs and vice versa. Moving is not for the faint hearted and we were to do it nineteen times in the twenty-seven years of our life together with the Army.

While we were still in Berlin we had used a leave period to return to England to look for boarding schools for Nick, because our various moves meant that his education was being badly disrupted. Not yet eight years old, he had attended kindergartens at Camberley, Londonderry and Shrivenham, had contended with vastly different curriculums between his first primary school in Australia and his second in Berlin, and now he was faced with the prospect of a new school in England. He needed much greater stability and boarding school was the answer. We did thorough research and came up with a short list of three – Haileybury Junior School in Windsor, a preparatory school near Hindhead, and another near Oxford. We visited both Haileybury and the Hindhead School and they impressed us, but the fees at the third were so much less expensive that we had to take a look.

We arrived to drive up a weed strewn approach to an old house with a grand façade. I had noted that the grass on the rugby pitches was over a foot long, which did not bode well. I pulled on a handle to ring a bell somewhere in the depths of the building. After a while the door was opened by an elderly gentleman who was the double of an old actor named Alastair Sim, who had starred as the headmistress in a 1954 film called *The Belles of St Trinians*. He introduced himself as the headmaster and that did not inspire confidence. He led us into a large entrance hall and up a sweeping staircase to his office. We saw no one else and the big house was silent. His office was spacious and once, long ago, its furniture would have been impressive, but now its Regency chairs were badly in need of re-upholstering and the whole room needed a good dust. His pitch was bumbling and entirely unconvincing. We asked if we could see some of the pupils, but he said that the school was on an outing. We thanked him for his time and left – hurriedly. Haileybury eventually got our vote and on delivering Nick there, he marched through its front door without even a backward glance. Upset Mum!!

Tim, also of school age, was very envious of his elder brother at boarding school. He had to make do with West Byfleet Primary School, which was not a success. Tim was a bright child and he quickly became bored because he found the teaching uninspiring and the general progress of his class too slow. We did what we could to encourage him at home, but he simply had to mark time until he was old enough to join Nick at Haileybury. On the upside, he had plenty of willing playmates on the quarters' patch.

Serving at the MOD meant commuting and, for the first time, travelling a significant distance to work. The commuter trains were old, crowded and frequently late, or cancelled. After a while I began to recognise people as fellow regular travellers, but most avoided eye contact and if a 'Good morning' was returned, it was done without any enthusiasm. I thought 'civvies' were a strange breed.

Gross overcrowding on the trains was the most galling inconvenience. Early on I elected to take a window seat, which was entirely the wrong thing to do. The train quickly filled at later stops and the other places on a bench seat of three were taken by two extremely large people. I was jammed, vice like, against the window. My train was halted by signals just outside Waterloo. Another commuter train drew alongside and came to a stop. It was an old, corridor train and its packed corridor was just a few feet from my window. Directly opposite me, with his cheek pressed against his window, was my boss, the Director of Army Staff Duties. We recognised each other and I was able to raise a hand to greet him. The general had more of a struggle to free one of his hands and, finally, he gave me repeated 'V' signs. He might have had two stars, but I had a seat!

After only a week or two, I was called to the Cabinet Office to be met by a senior official and taken to a briefing room where I was told that the information I was about to receive was Top Secret. It was combined with a code word, which gave it added protection. That meant that I could not discuss the subject matter with anyone, without the approval of specific Cabinet Office officials. The first area briefed was the government's survival in the event of nuclear war. I was the MOD staff officer responsible for ensuring that the principal bunker was kept maintained and fully operational. With the ending of the Cold War, the location of the bunker has been declassified and it is no longer operational, thus I am able to recount this detail. The second briefing concerned a different operation. It was as sensitive as the first and had its own code word protection. I was to hold the plan for this operation and update it as necessary. Such was the 'need to know' that my bosses were entirely unaware of my role and had no knowledge of these two operations. I had to hold the associated documentation especially secure and separate from all other plans. It was not for several years after I left the Army that I was debriefed and my vetting clearance was removed, presumably because the plan for the second operation had moved on. I am still unable to reveal what it involved.

The United Kingdom was in an unhappy state in 1981. The IRA mounted attacks in London and in Northern Ireland hunger strikes led to the eventual deaths of eight prisoners. Widescale coal pit closures were announced and then rescinded. Government spending cuts bit deeply and a newly formed Social Democratic Party overtook both the Labour and Conservative Parties in popular polls. Unemployment rose to above 2,500,000 and there were riots in Toxteth in Liverpool, Woolwich, Moss Side in Manchester and in Brixton. Women protesters established a peace camp at Greenham Common. Unrest, of one sort or another, seemed everywhere. All this meant that the UK desk was probably the busiest in ASD2.

The security situation in Northern Ireland was highly inflammatory. Troop levels required constant adjustment and every change to force levels needed governmental approval. That meant the writing of detailed briefs for ministers, the Chief of the General Staff and other very senior officers, as well as extremely close liaison with Headquarters United Kingdom Land Forces (HQ UKLF) and HQ BAOR, the military formations that provided all reinforcements to the Province.

While I was familiar with the way uniformed troops operated in Northern Ireland, I knew very little about covert organisations and their operations and wanted to learn more in order to be in an informed position to help staff their requirements at the MOD. I took up the offer of a two-day attachment to the SPG.

I was met at Aldergrove Airport by a SPG operative. We got into his unmarked car, but unlike the Mini that I had driven in Londonderry, this one had a radio – a very sophisticated radio. As we left the airport, my new colleague reported to his base that we were on our way. I could not see a microphone or the radio, nor had any antenna been visible when I had got into the car. The radio hardware was not so much concealed, as built into the fabric of the car. It was virtually undetectable. During our journey he occasionally reported progress by simply saying his radio call sign and then a two or three figure number. The numbers related to intersections, or cross roads in their area of operations – there were hundreds and all of them had to be learned by heart.

I was in countryside that I knew well – 8 Brigade's area. I recognised the barracks where the detachment was located, but had never been to the part occupied by the SPG detachment, which was shrouded from the rest of the barracks by a high, wriggly tin fence. The detachment commander began my briefing immediately. He knew that I had served in Headquarters 8 Brigade, and before that, with the Pompadours in the Creggan in 1973. He related some of his briefing to what had been going on in those times and told me things that I found difficult to believe. The whereabouts of some terrorists had been known, along with the locations of their weapons caches. I was amazed to learn that the SPG had been able to 'seed' some terrorist weapons with bugs, which would alert them if the weapons were moved and make them well placed to interdict an IRA operation. More often than not, the SPG would play the long game in order to put terrorists and weapons together, to make telling arrests that would lead to long convictions. The most remarkable story was of an operative who had been tasked to observe a meeting of Provisional IRA factions from Belfast and Londonderry that was held in the Bogside. He loitered outside the building where the meeting was to be held and both the Belfast and Londonderry factions took him as a member of the other group. As a result, he was ushered into the meeting. He stayed there for about half an hour, before indicating that he wanted to relieve himself and made good his exit. The detachment commander had ample evidence to support his stories. He then brought the briefing up to date and to conclude, he said that there was incomplete intelligence that terrorists were to move some weapons that night and the movements of the likely suspects were to be monitored. I could be part of the operation.

Probably six to eight cars were involved and the monitoring of suspects was nothing like the Hollywood version. Once a suspect was on the move, a tail vehicle would sit very well back, frequently out of sight of the target vehicle. Other cars would cover intersections ahead and would exchange tailing duties very frequently. Superb knowledge of the road and track network was needed for that arm's length tactic to succeed. After several fruitless hours we were content that no weapons had been moved and that the suspects who we had followed, had merely had nights out, independent of one another. Be that as it may, I had learned a lot.

Next morning, I was to get a feel of what it was like to operate, seemingly alone and in plain clothes, in the middle of a staunch IRA area. I was going for a walk in the Creggan! I was briefed and given a small pistol, but my principal weapons were a clipboard and a biro. The intention was to give me an idea how SPG operatives worked, frequently exposed and in the open. I was told that they wanted to update information about the occupancy, or lack of it, of a particular house. My part was to walk down Creggan Drive, a main arterial route that contained a row of shops, and make occasional notes on my clipboard as I went. That was the totality of my cover. I was looking for anything that indicated that someone was at home. Three cars would be orbiting me as I went.

I was dropped at a large roundabout in the north of the estate and then walked south towards the shops, stopping, as instructed, to make notes. I was surprised and delighted that passers-by took so little notice of me, but made sure that I did not make eye contact that might lead to conversation. I walked just beyond the target house. It had a full dustbin, but that could have been full for weeks. While it was inconclusive evidence, it was a piece for the intelligence jigsaw – if, on another occasion, the dustbin was empty, occupancy would be indicated. Just after I had crossed the intersection with Fanad Drive, a car pulled up beside me and my short, heart-stopping venture into the covert world was over. Until that moment I had seen nothing of the three orbiting cars and my eyeballs had been out on stalks looking for them. I had got the insight that I needed and that understanding would help me look after their interests. I took my hat off to the brave men and women of the SPG, quick to realise that covert operations were not for me.

Throughout the United Kingdom, industrial relations were volatile. MACM deployments were made in Great Britain to cover fire and ambulance strikes, involving personnel from all three armed forces. As those deployments were land based, the Army was the lead service and I had to work closely with my opposite numbers in the Navy and RAF to agree on the personnel, vehicles and equipment that they would contribute to any MACM response. Lesser deployments involved the laying of trackway in London parks to provide additional car parking while train strikes were in progress in the south east, and the removal of tons of rubbish from city streets when bin men went on strike. Preparations also had to be made for a possible deployment to distribute fuel to petrol stations when tanker drivers threatened industrial action. Tanker drivers had gone on strike in Northern Ireland and union officials there were encouraged by the military's failure to deliver sufficient fuel on day one of the strike, but by day three or four, all the Province's petrol stations were completely topped up and the strike quickly collapsed. Almost certainly that outcome led to the avoidance of industrial action on the mainland. In Great Britain the police had yet to become used to, or equipped, to deal with rioting and an urgent MACP request had to be rapidly met to provide riot gear to police dealing with serious disturbances in Liverpool. In winter, after blizzard conditions, some helicopter and over snow vehicles rescued people and animals from isolated communities in the Salisbury Plain area. I thought things could not get any busier, but 1982 was to be a year that stretched the Armed Services as no other since the Korean War, and the Army's implementation branch was right in the thick of it.

On 19 March, Argentinian scrap metal workers raised the Argentinian flag on the island of South Georgia, as a precursor to the invasion of the Falkland Islands. In ASD2 our first job was to find out exactly where the Falkland Islands and South Georgia were. We asked for maps and were told there were none available; for several days, a page torn from a Times atlas was our only reference.

Our offices were equipped with televisions and an early form of Ceefax to stay abreast of breaking news. I am not sure how it was achieved, but over that means we were able to listen to radio traffic from Government House in Stanley, capital of the Falklands, as it was besieged by Argentinian forces on 2 April. The broadcasts stopped abruptly after the Governor ordered the surrender of the islands. That same day, the cabinet approved the formation of a task force to re-take the Falklands, a decision reinforced, on the following day, in an emergency debate in the House of Commons. The MOD and all major military formations in the UK moved to a war footing.

The tri-service operations room opened in the MOD and the principal Army desk was staffed exclusively by officers drawn from the Military Operations and ASD branches. Initially, a single brigade was earmarked and was able to deploy rapidly (3 Commando Brigade), but another was quickly added (5th Infantry Brigade) and that would follow on as soon as it was able. While those brigades provided the bulk of the land forces, they needed an enormous amount of support to deploy with them. Our job was to identify all the forces required and to task the relevant headquarters to provide them. At the same time, it was vital that we keep our naval colleagues informed of the number of personnel, equipment and stores required to be moved in order that they find the requisite shipping. Ships were taken up from trade, which led to a very apposite acronym, 'STUFT'. It felt as if we were STUFFED at the outset, such was the volume of work. If I was lucky enough not to have an overnight shift in the operations room and had been able to clear the work on my desk during the long days in the office, I could go home and was often able to see the results of something I had worked on in the morning, on the 10 o'clock news that night. It was exhausting, but exciting and rewarding.

In the month that followed, SS *Canberra* and the *Queen Elizabeth 2* were requisitioned as troop carriers. In all sixty-two merchant vessels were STUFT, with helicopter decks welded onto many of them. The aircraft carriers *Invincible* and *Hermes* led the formidable naval component. As the elements of the land force were assembled and made ready for war, they departed south as soon as shipping became available. Some left within days of the decision being made to re-take the Falklands and by the first half of May, all 127 ships of the task force had departed. A period of waiting began as they sailed south and with strict radio silence imposed, only outline news about the progress being made reached us. It was a vulnerable time that would last until a firm beach head on the Falklands had been established. The operations room was a tense place to work.

As the task force neared its objective, its vulnerability to Argentinian air attack increased. I was beginning an overnight shift when a bell inside the teleprinter on my desk began to ring and a red light on its top blinked on and off. The same thing was happening to all the other teleprinters in the room. The bells and lights indicated a FLASH message. No one had ever received one before. FLASH would only be used for something like the commencement of hostilities. The teleprinters coughed out their message. It came from one of the aircraft carriers to Fleet – the naval headquarters. It was brief and read 'Exocet flying. Wait out'. An *Exocet* was a French made missile used by the Argentinian air force – one had sunk HMS *Sheffield* several days earlier. If a carrier was lost, along with the *Harrier* jet fighters it had on board, the whole operation would likely end in failure. The room went quiet as we waited for the follow up message.

That night, my clerk was an eighteen-year-old soldier from the WRAC. She was a first class typist, but had little other experience. She sensed something was wrong immediately. She asked what was happening and I explained the gravity of the situation to her. She said nothing for a moment or two and then said, 'I'm frightened.' She was an ASD2 clerk and was standing next to me. I knew her, so I squeezed her hand. As I did, the CGS walked in.

'What's going on?' I handed him the teleprinter message and he read it. He too was quiet for several moments and then said, 'Why were you holding that girl's hand?'

'She's frightened, sir,' I said.

He walked behind me and my clerk and briefly took her other hand. 'You're not the only one, young lady.'

General Bramall was a remarkable soldier, a Royal Academician and first class sportsman. He was a natural leader and I thought his compassion for a young, private soldier, in one of the most strategically telling moments of the Falklands war, was quite extraordinary.

South Georgia was recaptured on 25 April. Among the Argentinian prisoners was a Captain Astiz, an infamous person, wanted by France in connection with the disappearance of two French nuns during Argentina's internal 'Dirty War' of 1976–1983. The Royal Navy brought Astiz to Ascension Island, but during the voyage he allegedly attacked a naval officer. He was a really unpleasant and dangerous man and I was tasked to organise his onward movement to the UK and to find somewhere to put him.

When, in the previous year, the military had looked after civilian prisoners in Durham Jail, the man in charge there was the senior member of the Military Provost Staff Corps, which ran the Army's detention centre at Colchester. He was just the chap to look after Astiz and he was flown to Ascension Island to take charge of his prisoner. In the meantime, I had to find appropriate secure accommodation (the Geneva Convention precludes prisoners of war being held in civil prisons). My first thought was that the Secret Service would surely have some remote country houses that might serve the purpose, but MI5 and MI6 were entirely unhelpful. Astiz would have to be held on a military base where he could be kept separate from other personnel and away from prying eyes. The Red Fort at the RMP's depot at Chichester proved almost ideal, but it was not a prison and work was needed to make it secure. Royal Engineers saw to that in less than twenty-four hours and an artillery bombardier (corporal) installed an extensive network of CCTV cameras in much the same timeframe.

In the event, the UK government decided that it did not have grounds to prosecute Astiz as his alleged crimes had taken place in Argentina and were not then defined as offences under international law. After only a few days in the Red Fort, he was repatriated. He was flown by RAF helicopter from Chichester to the end of a runway at Gatwick Airport, where he and his minder boarded a civil airliner that had its engines running. The dramatic transfer between the two aircraft was shown on national news that evening. Subsequently, he was handed over in Montevideo, to Uruguayan authorities, before being returned to Argentina. Astiz was discharged from the military in 1998 and his past eventually caught up with him. In 2011 he was sentenced to life imprisonment for murders that included those of the French nuns.

At last, the task force arrived at the Falklands. In brief, 3 Commando Brigade landed at San Carlos Water on 21 May to establish a beach head. From there forces moved south to fight the first of the land battles at Goose Green on 27–28 May. With Argentinian forces removed from Goose Green, the way was open for units of 3 Commando Brigade to fight eastwards towards the capital, Stanley, on foot and by helicopter. The second brigade, 5 Infantry Brigade, arrived on 1 June and with those reinforcements landed, the assault on Stanley could begin. The Argentinians held the high ground around the town and intense battles were fought over the period 11–14 June to gain those dominating features. When they fell, the Argentinian commander was forced to surrender.

There was enormous pride throughout the country in what the Armed Forces had achieved, as well as a deeply held sense of relief. In the MOD much of the tension dissipated and some sense of normality returned. Meanwhile, the troops in the south Atlantic took stock of themselves, moved to assist the Falklands' civil population, and worked out how best to look after thousands of Argentinian prisoners. Out of nowhere I was to be involved with the latter.

One Tuesday I was at my desk when Colonel Richard Swinburn, who had replaced Colonel Tony Makepiece-Warne as my immediate boss, walked into the office to say, 'Colin, do you know how to handle POWs?'

'Well, I know how to take one safely. You get taught that in basic training.'

'You'll do,' he said.

Colonel Richard explained that the Prime Minister had expressed a wish to return all of the Argentinian prisoners to the UK. 'You're joking,' was all I could say.

'Don't worry. The idea's hair brained and will die a death before much time and energy are spent on it.'

My first task was to identify somewhere to put them. Within hours of being given the task I received a telephone call from a Red Cross official in Geneva asking for reassurance that the POWs would be accommodated according to the rules of the Geneva Convention. How he got to hear of the idea and found my name and telephone number is still a complete mystery. The call made me take my task increasingly seriously and I used my contacts in HQ UKLF to help.

They too thought it was a joke, but someone, much more senior than me, told them to get on with it. A couple of days went by and progress was made, but there was still no sign of the idea being discounted. HQ UKLF then came up with the possibility of requisitioning accommodation at Sullom Voe, which was then serving as British Petroleum's oil terminal in the Shetland's. It would have to be made secure and additional facilities would have to be built, but that work could probably be done in the time it would take for the task forces' ships to make the return journey to the UK.

It was now Friday and I was instructed to write a paper for the Chief of the Defence Staff (CDS) to take to cabinet early on Monday morning. The staff at HQ UKLF undertook to provide me with more detail to support my paper by first thing Sunday morning. I went home to make as much of Saturday as I could. I had the main office to myself on Sunday. My clerk, Sergeant Bevis, brought in the information sent by HQ UKLF and I settled down to write. In those days we wrote in long hand and our clerks typed our script. Sergeant Bevis took each page as I finished it and kept me supplied with tea. I completed a first draft by about ten on Sunday evening. Proofreading and 'cut and paste' took another few hours and Sergeant Bevis had produced an error free copy by around two on Monday morning. Just before four am Colonel Richard Swinburn arrived and took my paper to his office to read. He came back about an hour later. He had made only a few comments and said, 'Well done. It reads well. Freshen up and get some breakfast and take the paper to the CGS at 7 o'clock.'

I did as I was told and was shown into General Bramall's office. He read the paper and said 'Good paper. You know more about it than anybody else, so you'd better take it to the CDS.'

A few minutes later I was in front of the head of the Armed Forces. Admiral Sir Terrence Lewin took my paper and read it carefully. Later that morning he would present it at cabinet. He put the last page down on his desk and looked across at me.

'This might work,' he said.

'Yes, sir,' I replied.

'Well, I bloody well didn't want it to!' Nobody had told me!

'Not your problem,' said CDS.

Subsequently, he was successful in making his case to the Prime Minister to return the opposing forces to their own country. The scheme was dropped and disappeared without trace. After that, my work on the Falklands War wound down and normality was restored in ASD2 well before the victory parade was held in London, in October 1982.

While I was busily engaged with fluctuating force levels in Northern Ireland, responding to requests to employ troops to mitigate the effects of industrial action, and doing my bit for the Falklands war, Sue made sure that family life did not suffer unduly. Nick had settled well at Haileybury Junior School and because it was only a thirty-minute drive to Windsor, he came home for a few hours most Sundays, after he had been to church – he was in the school's choir. Tim had made lots of friends and, outside of school, was happy. Sue found part-time employment as a market researcher and enjoyed the variation that the research projects brought – anything from public reaction to the planned route for the M25, to preferences for brands of cigarettes. Family holidays were important and in 1981, ours had been to camp in Brittany for two weeks. We were blessed with wonderful weather until the last night, when we discovered that our tent leaked, making for a slightly damp journey home.

In the following year, once hostilities had been brought to an end on the Falklands Islands, there was a rush to take leave and draw breath. Sue and I approached our great friend, Paul McMillen and we decided to take both families to holiday in Scotland. For two weeks, we hired a farmhouse, perched on the side of a steep valley, a few miles outside Moffat, in Dumfries and Galloway. Paul had two children, Sarah and Emma, with his first wife, but the marriage went disastrously wrong when she left both him and the children. Paul brought his girls up magnificently well as a single dad, until he found Alison over a decade later. He was Nick's godfather; Sarah is my goddaughter, so the holiday was a great opportunity for family bonding and to get to know Alison better. Paul died, all too prematurely, in 2012. He was a second brother to me. I miss him tremendously.

Around the middle of 1982, the 'Pink List' was published. It contained the names of those officers who were to be promoted to lieutenant colonel. Mine was on the list and at the earliest opportunity. That was not only satisfying, but important because it indicated that I would be in the running for a command appointment. A few weeks later, the command list was promulgated and I had been selected to command the Pompadours. I was absolutely over the moon. I rarely thought beyond my next appointment, but command of one's own battalion is the dream of every professional infantry officer. Wow! Wow! Wow!

There was a downside. The next appointments for officers on the 'Pink List' would be as lieutenant colonels. Given that all of them were part way through majors' appointments, most would be promoted within a year, some within months. In my case, the command list changed that expectation because the current commanding officer of the Pompadours, Alan Thompson, was only a year into his appointment, which would run for some thirty months. I would have to wait until December 1983 for my turn to command.

I was contacted by the Regimental Secretary, a retired officer who coordinated officer postings. The current second-in-command of the Pompadours was due to be posted in January 1983 and the regiment had no suitable replacement. It was proposed that I become Alan Thompson's second-in-command for several months, until a replacement could be found. I argued strongly against it. If I did as suggested there would be two bosses in the battalion. Inevitably, people would have one eye on Alan and the other on me, as his nominated successor. I regarded it as a most unhealthy solution and one to be avoided. My arguments did not prevail and both Alan and I had to accept the situation. I would join the battalion in Colchester in January.

The last few months at the MOD were unremarkable. My job had been interesting and I had done well, but I heartily disliked the political manoeuvrings that seemed endemic in government departments and I hated commuting. For much of the time, I felt like a fish out of water and longed to be charging about the countryside, with some good blokes and a rifle in my hands. Colonel Richard Swinburn was aware of my feelings and wrote in my annual confidential report 'Major Groves so loathes the MOD that he should be spared serving there again'. I cannot be sure how much efficacy that comment had and did not know it at the time, but when I skipped down the steps of the Main Building, just before Christmas 1982, I was leaving the MOD never to return. Hooray!

MARKING TIME FOR COMMAND

In the early days of television, when there were breaks in transmission, 'INTERLUDE' would appear on the screen. I feel the same way about 1983. I recount this period as an appendage to my time at the MOD because that is exactly how it felt. I spent the year anticipating the moment when I would take command of my battalion.

After Christmas in West Byfleet, the first big event of the New Year was for Sue and I to take both our sons to Haileybury Junior School. Tim would not have his eighth birthday until March, but the school recognised that with a family move impending, it would be better for him to go to boarding school a little early, rather than have him spend just one term at another school. He was delighted to be going and following his brother's lead, he too walked into the school without looking back. 'Groves 1' and 'Groves 2' were in residence.

Sue and I completed the usual round of packing and cleaning before handing over our quarter and a removal van transferred our effects to Colchester in mid-January. As a nominated commanding officer, our quarter at 24 Lees Road was much more generous than the West Byfleet semi and we enjoyed its space.

I re-joined the Pompadours in Meeanee Barracks and immediately began taking over from the incumbent second-in-command, Tony Taylor. The handover was done quickly and efficiently and there was work to be done immediately. During my first conversation with Alan Thompson, he told me that the battalion had been warned that it was to one of the first to train in America, at Fort Lewis, in Washington State and at a field firing area relatively nearby, at Yakima. The second-in-command's principal duty is the co-ordination of training and I was to accompany Alan on a joint reconnaissance with representatives from a Ghurkha battalion, the other battalion to train at Fort Lewis that year.

Even before departing on the recce, I had noticed a change in the way the senior personalities in the battalion regarded me. Although I was still 'Colin' to them, they were sensitive to the fact that in only a short time, I would become 'Colonel', or 'Sir'. The distancing it brought was subtle, but it was there nevertheless. What I had feared, when I had first heard of my second-in-command's appointment, was happening as people recognised that there was a boss and a boss in waiting.

Fort Lewis is some 40 miles north of Seattle and it was then home to 1 (US) Corps. Like most American bases it was huge, certainly as large as a medium-sized town. The impending arrival of British troops had stirred some interest and both we and the Ghurkhas were under a heavy remit to make a favourable impression, in order to secure the prize of continued access to the training facilities in Washington State. The recce parties had arrived at the weekend, but the Range Officer at Fort Lewis, Mr John Weller, a former US Army artillery officer, opened the Range Control building on Sunday so that work could get off to a flying start.

John Weller made an immediate impression. He was a man on top of his job, with a ready, laconic sense of humour. His briefing of what he could offer by way of training facilities, exceeded our expectations. He asked if he could stay with us while we began to design the training. The recce parties did this separately and within a couple of hours, both had a comprehensive outline of what they wanted to achieve and what facilities they needed. John Weller was surprised at the autonomy British battalions were allowed, how quickly the training had been designed and the bids for his facilities had been submitted. Apparently, in just a morning, we had achieved what would have taken weeks under the American system, where high level approval was required before training could be sanctioned. We had made a staunch ally in him and I had made a lasting friend. Sue and I are still in close touch with John and his wife, Edie – they made a significant contribution to our 50th wedding anniversary celebration during a visit to New York in 2019.

The recce parties returned to England, but I was not to be there for long. I was back in Fort Lewis, with an advance party, well before the main body of the battalion arrived in early March. The advance party had two main tasks to prepare for the arrival of some 650 soldiers. Mine was to flesh out the outline training package that we had designed and to write detailed instructions for individual exercises. The quartermasters had to take over accommodation and make arrangements for the provision of rations, transport and a myriad of other domestic details.

A third task became evident – to uncover differences in operating that might cause for friction, misunderstanding or confusion. The first was that the Union Jack could not be flown on a flag pole higher than any in the vicinity flying the Stars and Stripes – reasonable enough. The next to be discovered was policing. American military police brook no nonsense to the point that they are remarkably heavy handed by British standards. Pistols seemed to be drawn as a matter of course and that would go down extremely badly with our boys. We immediately made arrangements for the main body to come with two RMP NCOs and they would work with their American counterparts to intervene in instances where British soldiers were involved. The last anomaly was accounting. The advance party had no transport of its own and as Fort Lewis was so big, that posed a real problem. Our senior quartermaster, Stan Bullock, had found a friend in a US supply organisation and asked if he could borrow a Humvee vehicle on the old boy net. Stan was surprised to be refused, but the American officer said, 'I can let you have five!' Stan had no use for five, but it was explained that Humvees were parked in mini rows of five vehicles. If one was removed, it would be obvious. A row of five would not be missed. Consequently, we had an embarrassment of riches and an early appreciation that, logistically, Americans think big.

I designed the training in three phases to last a total of five weeks. The first, at Fort Lewis, made use of conventional ranges and some add-ons not available in the United Kingdom, like the Tyre House – a roofless building, made of earth-filled old tyres that absorbed live rounds and fragments from grenades, to permit extremely realistic house clearing drills. The next phase saw the battalion move to Yakima, to the live firing area. Our hosts could not have been more generous in the support of our training there. I was astonished to be asked if I would like to fly the battalion to and from Yakima. Of course, I said 'Yes' and C130 transport aircraft turned up, flown by the National Guard, who were happy to be involved.

When Brits get access to large field firing areas, they are the ones to think big – the British Army is renowned for its expertise in live firing exercises that come as close to the realities of battle as is possible, without risking safety entirely. I asked if the Americans would be interested in providing artillery support for certain of our exercises. By this time, we had captured their interest and a battery (six guns) was allocated to add to our own mortars and medium machine guns. The cherry on the cake was that the American Air Force wanted a piece of the action and fighter ground attack aircraft were written in the exercises as well.

After returning to Fort Lewis from Yakima, some soldiers were able to make use of the Huckleberry Creek Winter Warfare School; some intrepid souls climbed the nearby volcano, Mount Rainier; and as a climax to our training, we exercised against the 2/75 Ranger Battalion (a special forces unit), to fight a diplomatic draw.

Interspersed with all this action man stuff, we 'flew the flag'. The Band and

Drums 'Beat the Retreat' in front of a large audience of 1 Corps families; the officers held a cocktail party for 200 guests – we had brought a selection of our silver with us and that displayed with the Colours, created quite an impression; a series of lunches were held to thank the many Americans who had helped us; and, as a parting farewell gesture, the battalion paraded for the Commander of 1 Corps, to show off the best of British drill. As a battalion, we could not have done more and Fort Lewis continued to host two British battalions annually for the next twenty years.

During my remaining months as second-in-command, the Pompadours collaborated with the five other battalions of The Royal Anglian Regiment, to host a visit by the Queen Mother; separately, they took on the Spearhead battalion commitment and were involved in three major exercises. While all that was going on, conversion training began in order to be ready to assume the mechanised role (i.e. be equipped with APCs), when the battalion moved to Minden, West Germany at the turn of the year. The mechanised role is complicated and involved hundreds of soldiers attending dozens of different courses, with the programme of training running for seven months.

I could not serve as second-in-command beyond mid-summer. As a commanding officer designate, I had my own courses to attend and it would have been increasingly difficult to have me in daily evidence, right up to the time I took over from Alan Thompson.

My successor as second-in-command would only spend a few months with Alan and, thereafter, he would support me, so the next incumbent was my choice. I chose Peter Dixon, then commanding the Pompadours' C Company. He had carried the Colours with me at Hertford in 1969 and the more I saw of him, the more certain I became that he would be exactly the friend and advisor whom I would need. Peter took over from me and I went on 'gardening leave'.

For the next four months I had a generally idle time. I was in purdah and avoided all contact with the battalion. I was required to attend a commanding officers' course and to command a small armoured battlegroup – a force numbering about 900, consisting of tanks and APC equipped infantry, plus artillery, engineers, helicopters and logistical support – on an exercise on Salisbury Plain. As a commanding officer designate of a mechanised unit, it gave me a chance to experience what was coming my way.

Before an officer relinquishes, or takes command, they are privileged to 'kiss hands' with their regiment's royal Colonel-in-Chief. As The Royal Anglian Regiment was a large regiment, each of its battalions had its own deputy colonel-in-chief and on 17 December, Alan Thompson and I had an audience with Princess Margaret at Kensington Palace. In order that she could devote more time to us, her equerry thoughtfully put us at the end of a long line of people who she received that morning. The audiences overran badly and by the time we were shown into a charmingly decorated, but lived-in drawing room, Her Royal Highness had had enough. Her first words were to her equerry – 'I could do with a drink' and as an afterthought 'Give these gentlemen one too'. Alan and I found ourselves standing in front of a tiny woman, clutching large glasses of gin and tonic. We had a pleasant, relaxed conversation, but it was cut short by a dull explosion. Only a mile away, the IRA had detonated a car bomb outside Harrods that killed six people and injured ninety. It was a dreadful way for my jinx on royal visits to continue.

On the battalion's last working day in Colchester, I was able to put on the badges of rank of a lieutenant colonel. A staff car collected me from my quarter and I was driven to Meeanee Barracks. Alan Thompson stood outside battalion headquarters and we were joined by the commanding officer of 1st Battalion, The King's Own Scottish Borderers. The three of us shook hands. The Borderers were relieving the Pompadours in Colchester and Meeanee was their barracks now. I was assuming command and Alan was off to his next posting in Zimbabwe. I left immediately, leaving Alan to enjoy his last moments as CO and the traditional farewell of being pulled out of barracks.

The family spent Christmas in our quarter at Lees Road and we had a full house on Christmas Day, when we were joined by my mother and my favourite uncle, Peter, who lived in Colchester. When the festivities were over, we prepared for the move and the handover of our quarter. We, and about 1,200 soldiers, wives and children of the 3rd Battalion, moved to Minden in the first days of the New Year.

CHAPTER 14
COMMANDING THE POMPADOURS

PART ONE: MY FIRST YEAR

Bitterly cold weather greeted our arrival in Minden, a town that straddles the River Weser in the centre of the north German plain. In 1984 it had a population of some 70,000 people that included a large garrison, with both British and Bundeswehr units stationed there. The Pompadours took over from 1st Battalion, Royal Regiment of Fusiliers to become part of 11th Armoured Brigade. The Fusiliers were commanded by Pat Shervington, originally a Royal Anglian and a fellow subaltern with me in Berlin in the 1960s. Our long friendship made for an easy handover. Always the practical joker, Pat had hidden forty-four fusilier hackles in our new quarter – a hackle is a feathered cockade that fusiliers wear in their berets, behind their cap badges. The number forty-four had significance because one of the Pompadours forebearer regiments was the 44th of Foot. For weeks after we had taken over our comfortable house, Sue and I continued to find those things in the most obscure places. Months later, we had found forty-three and I was convinced that Pat was playing mind games, hoping that we would spend the rest of our time in Minden looking for the forty-fourth, which probably never existed, or more likely, Pat couldn't count!

The battalion's new home was Elizabeth Barracks. It was located on the eastern side of the Weser, while the rest of the garrison and the town's major amenities were west of the river. It isolated us to a degree. The soldiers' accommodation and the various offices and stores were all in old Kaiser-style barrack blocks. A few newer buildings included the NAAFI canteen and small shop, the officers' mess and garages. The garages posed a big problem because there were nowhere near enough, which meant that most of the APCs were parked in the open throughout the year, around the edge of the parade square. The next problem was with the APCs themselves. They were petrol driven Mark 1 432s, much older than the majority of my soldiers, in fact my Technical Quartermaster, John Rourke, recognised one of them as the vehicle he had driven more than twenty-five years previously when he was a private soldier.

I was told that they were so old they were eligible to take part in vintage car rallies! Keeping them motoring would be a never ending task.

With that maintenance problem firmly in mind, I had liaised closely with Captain Dick Martin, who commanded the Light Aid Detachment (LAD), attached from the REME. The LAD was responsible for the in-house repair of the battalion's weapons, vehicles and technical equipment. They were an integral part of the battalion and identified with us to the extent that they wore our brown beret, while retaining their own REME cap badge. Dick made clear just how much expertise my soldiers would have to develop, over and above the basics taught on the various conversion courses held during the previous year in the UK. With his advice firmly in mind, I designed a short training exercise for each of the companies, to take place at Haltern, a training area quite some distance from Minden. Each company would deploy in turn and learn what was involved in moving their APCs by rail; get used to living in and from them; practise some tactical drills; and master day and night refuelling. Most importantly, they would have to maintain their vehicles in adverse weather conditions and that would involve daily routine checks, breaking track (taking the track off a vehicle, normally to replace damaged links); assist their REME section with pack lifts (engine replacements) and with recovery of broken down vehicles. It was an extensive menu and one to keep them fully occupied.

I was disconcerted by rumours that quite a number of soldiers' quarters had been left in poor condition. I spoke to Bill Burford, the Families' Officer, to see what could be done. We decided that the best thing would be to hold a wives' meeting that I would chair, during which the wives could vent their grievances publicly. The meeting was held in the soldiers' dining hall. There were about 250 wives in the Pompadour family and if half of them turned out that would be a good result. In the event, nearly all of them were there. They presented a formidable audience.

From the start, it was clear that barely a handful of quarters had required a little minor cleaning on handover. Just three had caused a significant problem. Once that fact was in the open, it dispelled the rumours. I went on to ask how many had something they wanted doing in their quarter, be it maintenance, or the replacement of furniture or fittings. Nearly 250 hands went up! We did a deal. If the wives would write down their name, their quarter's address and the work, or replacement action, they required to be done, then the battalion headquarters' clerks would type, what turned out to be, more than 200 hundred forms that would put action in hand. The plan worked brilliantly and in coming months, virtually every request was met. The wives were happy for the support they received, but they had other issues that they wanted addressed.

First, they wanted a nursery, a bigger, more attractive one than the battalion had organised in Colchester and we delivered it within weeks of arrival. Next, they wanted help in finding jobs. That was not a problem anybody could solve, but it was a major penalty for wives accompanying their husbands aboard. In Colchester, 144 wives had either full time, or part time, employment. On arrival in Minden, perhaps only half a dozen had found jobs. That number would not grow significantly, primarily because of the language barrier, which meant that for the vast majority, employment was limited to that generated within the garrison. Serving in Germany attracted a local overseas allowance, but it did not compensate for the loss of income for many wives who had lost their employment.

Notwithstanding the unsolvable employment problem, I recognised that the wives valued the commanding officer speaking to them directly and having the opportunity to have their queries and concerns answered by me, with no intermediary putting their gloss on the issues in hand. I undertook to brief them a couple of times a year. The wives loved that. Some of their husbands were less enthusiastic because, for whatever reason, they were frugal with the information they passed on. Interestingly, while abroad, commanding officers also had legal command of their unit's wives and children. All dependants were subject to military law and that device enabled the West German authorities to turn over to British jurisdiction any case they did not wish to deal with themselves. A good example was that during our tour and elsewhere in Germany, a British wife murdered her soldier husband. She was tried by general court martial (an assize court equivalent), found guilty and subsequently, served her sentence in a British prison.

It felt like a lot had been achieved in a couple of months. Alan Thompson had handed me an efficient unit. I found him to be somewhat dour and the battalion tended to reflect his mood. The Pompadours were competent and confident of their abilities, but they were not a very smiley bunch. They lacked the sparkle that a happy, professional outfit should have. Putting a spring in the battalion's collective step was my next goal.

Peter Dixon was already proving an inspired choice as second-in-command. While I came up with the direction and outline for training, he put the flesh on the bones and made things work. He was also a wonderful sounding board on any matter. My quartermasters, Stan Bullock and John Rourke, were exceptional and the company commanders varied from very good to excellent – two were later to command the battalion. I also had the brightest bunch of subalterns that the Pompadours had known since my own generation in the early 1960s. One was to become a lieutenant general; another a major general; two became brigadiers; one more would have easily attained that rank, but chose to leave as a colonel, and the remainder all had successful careers. On top of all that I had an extremely powerful and proficient warrant officers' and sergeants' mess, led by RSM Bob Eke. Arguably, the Pompadours major strength, over decades, was the closeness of the relationship between its officers' and sergeants' messes. The warrant officers and sergeants took care of daily routine and they educated the young officers in man management skills and tactical nous, the sort of things that are not to be found in books and only experience can teach. It was a great and lasting partnership.

By a stroke of good fortune, the divisional commander (a major general) had announced that a divisional inter-platoon competition would take place in late May, only four months after the battalion's arrival in West Germany. Five battalions would take part, each sending three rifle platoons. As 'newbies' we were not expected to do well.

Our early training with the APCs would help, but I recognised that APC mounted manoeuvres and maintenance would be our nemeses – we were still firmly in the learning stage. Fitness and shooting always provide a firm foundation and I had no concerns about those aspects. Tactical proficiencies, when operating on foot, were generally sound, but other skills like first aid, radio procedure, night navigation, nuclear, biological and chemical (NBC) drills and armoured vehicle recognition – including those of the potential enemy and allies – would also be tested. Company commanders designed their own work-up training for their platoons and the training culminated with the battalion's inter-platoon competition. As luck would have it, each of the three rifle companies had a platoon in the top three places and all would be represented in the divisional competition.

The competition was held over three days at the Sennelager Training Centre. After the first phase, which focused on APC mounted operations, the Pompadour platoons were firmly in the lower half of the pack. My brigade commander, Brigadier Jeremy Blacker, a Royal Tank Regiment officer, kindly told me that I should not be too upset if that was where they remained. I surprised him by saying that I thought that was most unlikely. A night shoot and night orienteering exercise saw our situation improve and our platoons moved to mid-table. The various skills tests improved their positions still further. The final morning brought a long, forced march and the culmination came with a shooting competition that gave no time for the platoons to recover from their exhausting trek. Tired men's fitness and shooting abilities would determine the final result. We did not win the competition, but our platoons came second, third and fourth and we were easily the highest placed battalion. The Pompadours had arrived in some style – and my boys were smiley.

That smile broadened in the next few weeks when the battalion's two teams excelled in a BAOR competition called the *Connaught Shield*. It involved NBC skills, armoured vehicle recognition, shooting, and most particularly, first aid and casualty evacuation. The teams fought their way through earlier rounds to qualify for the BAOR finals. The casualty evacuation phase was the killer. A soldier casualty had to be carried on a stretcher for several kilometres, over rough terrain that included a bog and a waist deep pond. Only the fittest soldiers would be up to its demands. Our teams came 2nd and 4th. All ranks now expected to do well in whatever they undertook and I was immensely proud of all they had achieved in the first few months in Minden.

From a professional viewpoint, the first year in West Germany was a particularly demanding one. Not only did the battalion have to master its new mechanised infantry role, it had to learn to operate as part of a combined arms battlegroup. The Pompadours were twinned with 3rd Royal Tank Regiment (3 RTR) and I sent my 'B' Company to join them as the infantry component of their battlegroup. I received the tanks of 'C' Squadron 3 RTR in return. Whenever we deployed for a major training exercise, our associated tanks, artillery battery and engineer troop came with us and the exercises came thick and fast.

By early summer the battalion had undertaken its initial training as a battlegroup at the Soltau training area. That was followed by a two-week field firing period at Sennelager, when all the battalion's weapons were fired in tactical settings. Following that there was a pause until the farmers brought the harvest in, which allowed big formation training exercises to take place over open countryside. No exercise was bigger than *Exercise Lionheart*. It involved all of 1st British Corps, including the components from the UK, which would join it on mobilisation. About 100,000 troops would be involved on the British side, plus a division of the Bundeswehr, roughly another 20,000, who would provide the 'enemy'. As an immediate work up for *Lionheart*, the battlegroup spent another two weeks manoeuvring at Soltau and deployed from there directly to its start positions for the Corps exercise.

The exercise was a practise for war, over ground that would see the battles of World War III should the unthinkable ever happen. The battalion spent the first days preparing a defensive position. Soldiers dug their fire trenches and were further protected by dummy minefields and barbed wire obstacles. The work was done wearing NBC suits, with respirators kept handy, because the scenario painted the possibility of the early use of nuclear and/or chemical weapons. The debilitating effect of wearing the suits made hard work even harder. Once the defensive preparations were complete, the Corps sat tight for a couple of days while air reconnaissance was flown against it, trying to determine its positions, which were heavily camouflaged. No radio traffic was permitted in order to prevent the enemy's electronic warfare assets from pinpointing our positions. Only when that phoney war phase was complete were battles fought, with umpires determining the outcome of the many engagements. After three days of mock fighting, there was a chance to redeploy before the West German weekend enforced a mandatory pause.

I was made aware of just how big my battlegroup was when, late on a Friday night, we had to move to a new position. In preparation, the infantry companies, tank squadron, artillery battery, engineers and logistic elements had all positioned themselves in separate convoys and were parked, in darkness, on the side of roads. When the order to move was given, all the vehicles turned on their side lights and the convoys rumbled off. It was as if Christmas had come early as long, long chains of light wavered over the otherwise inky black countryside. There was no mistaking that there was a lot of ironmongery out there.

The exercise's next phase saw the Corps practise offensive operations to drive the enemy back from gains made in the first week. When that was all over, the clearing up began. The defensive positions were filled in. The dummy minefields and wire obstacles were recovered and compensation was assessed and paid for the inevitable damage to fields and roads.

But for the Pompadours, the exercise season was not complete. We had been tasked to provide the headquarters and infantry components for a high profile tri-partite exercise.

Throughout the Cold War, most people in the UK probably felt somewhat distant from it. Certainly, they were unaware of the detailed preparations that were made in West Germany. One of those envisaged the Soviets closing the autobahn to Berlin if relations between NATO countries and those of the Warsaw Pact deteriorated drastically. A plan was practiced annually and *Exercise Treaty* involved forces from Britain, the USA and France – the occupying powers in West Berlin. They came together to form a battlegroup that would cross into the DDR to travel along the autobahn to Berlin. That was their legal right. Armed resistance by the Warsaw Pact to the battlegroup's progress would almost certainly herald World War III. Because the allied battlegroup would assemble and leave from the part of West Germany where the British were stationed, it would have a British commander – me!

The exercise took place in November. For months beforehand, exchange visits took place between units of the three nations and a study day had been held. The British provided the headquarters, two infantry companies, the recce, mortar and anti-tank platoons and some reconnaissance helicopters. The Americans provided a company of M1 tanks and engineers and the French were represented by two artillery batteries. All three countries brought their own signal and logistic elements and the battlegroup had fighter ground attack aircraft support from the three national air forces. It was an extremely potent battlegroup, but one with no hope of success if it were to be violently opposed by the huge armies that would stand in front of it. The plan amounted to a political last throw of the dice. To that end the signal elements were vitally important. Before the exercise could begin, the signallers had to establish relays to make contact with the offices of two presidents and one prime minister. The heads of government played no part in the exercise, but I was told that they were aware of it and officials in their offices gave permission for the exercise to proceed.

The battlegroup formed up on Sennelager Training Area and an adjoining major arterial route was used to represent the autobahn to Berlin. For two days the vehicles of the battlegroup moved and stretched up to 20 kilometres along the route, sometimes moving off it to avoid simulated Warsaw Pact air attack, and all the time civilian traffic continued to flow as best as it was able. To conclude the exercise, the battlegroup returned to Sennelager so that some tactical manoeuvring and battle procedures could be practised – not easy when some American procedures differed from our own and the French, then not even a part of NATO, were completely different. They did their own thing, but it all came together in the end and everyone went home pleased with the outcomes.

The major exercises were only a part of the training schedule. A whole host of other training events took place. Various cadres were held within the battalion – some qualified private soldiers for promotion to lance corporal; others fitted soldiers to become signallers, mortar men, anti-tank missile numbers, drivers and assault pioneers (in-house engineers). Dozens of external specialist courses covered anything from training as a medic, to a storeman. Long, professional courses included those at the School of Infantry to qualify corporals as section commanders; sergeants as platoon sergeants; and young officers as platoon commanders. That list barely scratches the surface and the totality of training was impressive. With external courses excluded, the battalion spent nearly 200 days training out of barracks in its first year in Minden. But if life was to be lived at that pace, then it was essential that some slow laps had to be run.

The divisional inter-platoon competition had raised morale and put a smile on our collective face. For me, the trick was to keep things going. Winning, or at least competing well, was one way of doing it. Sport provided part of the solution. In the first half of the year, hockey, athletics and swimming were our strengths. The hockey team became overall Infantry champions. The swimming team were divisional champions, runners-up in BAOR and 4th in the Army championships. Athletics was a work in progress, but nevertheless our squad qualified to take part in the BAOR finals. Football and rugby were enjoyed and played at a reasonable level, but with no great success and boxing was to star later in the year. Professional sensibilities would not allow shooting to be regarded as sport, but it was certainly competitive and the battalion's shooting team qualified for the Army championships held at Bisley. So far, so good.

Many of the young soldiers and their equally young wives had not lived outside of the UK before and most missed their homes. Leave was vitally important and I made it a priority that every soldier got his full entitlement. It was a difficult process because no unit was allowed to fall below a minimum manning strength and the number of soldiers away at any one time was strictly governed. Trickle leave – having a relatively few soldiers on leave at any one time – was the answer and it took place over most of the summer months. I accepted that for everyone to get their leave, a number of key personnel would be missing for some training periods. The downsides were that they missed the training experience and the battalion was temporarily deprived of their expertise. The upside was that they came back refreshed and raring to go and, undoubtedly, their reinvigorated effort repaid me handsomely. The wives were particularly grateful to have that part of their domestic lives guaranteed to them. For our part, Sue, the boys and I drove to Austria to holiday in the Alps.

Schooling is another vitally important aspect of family life and the education of Service children was organised centrally in BAOR. The battalion had what amounted to its own primary school. It was very conveniently sandwiched between our barracks and the married soldiers' quarters' area. Just before Christmas 1983, the school was populated, almost entirely, by the Fusiliers' Geordie children. By mid-January, after the changeover of battalions, the vast majority of its pupils came from Bedfordshire, Hertfordshire and Essex – an enormous transformation. I was impressed by the seamless way the teaching staff had absorbed our children and settled them quickly. Secondary education was provided at the Prince Rupert School at Rinteln, about 15 miles from Minden. It was a comprehensive school with a wide catchment area and a considerable turnover amongst its student body, caused by their parents' postings. Small wonder that today Service children qualify for Pupil Premium funding because of the disruption to their education.

Traditionally, the commanding officer's wife organises the Wives Club. Sue recognised the value of the club, but her experience was that officers' wives and those of sergeant majors tended to provide virtually all of the impetus, to the unwitting exclusion of younger, less experienced wives. She thought that the pool of volunteer helpers would be widened if activities were organised at a lower level and companies organised their own events, but only if they were prepared to do so and a demand was evident. She would provide whatever support was necessary, but that would be the limit of her involvement. There was some fluttering in senior dovecots at this departure from the norm, but the system worked well and the various events probably attracted a better attendance than had previously been the case. Sue put her energies elsewhere and she became the librarian at the Prince Rupert School – a qualified librarian was a rarity and a considerable bonus for the school.

My last effort to keep the battalion happy in its first year in Minden, was to have a families' open day organised. My briefings to the wives had revealed how little many wives knew about what went on in barracks and what happened when the men were away training. The families' day was a way to correct that and to have some fun. The morning was spent with wives and children allowed to crawl over virtually all of our weapons, vehicles and equipment. Their keen interest was self-evident and they made the most of the opportunity. The cooks excelled in providing a lunchtime barbeque for over a thousand. The afternoon was taken up with 'It's a Knockout' competition, where everyone took part, if only to cheer their teams on. I have no idea what the evening held and it is probably as well that I chose ignorance, but I was certain that I had a happy bunch. That certainty was reinforced over the Christmas period when some twenty-one parties were held by the various components that make up a battalion. Sue and I were invited to most of them and there was no mistaking the upbeat mood that prevailed everywhere we went. The round of parties was fun, but it was exhausting. A night off was something to be treasured.

In the weeks before and after the New Year there was a big change to my command team. Peter Dixon had left as second-in-command to be replaced by Julian Lacey, a studious, able and, unusually, a chain smoker, who had big shoes to fill. Three company commanders had come to the end of their tours and their successors were in post; Captain David Clements was replaced as adjutant by Captain Frank Froud; Bob Eke was commissioned and was succeeded as RSM by Sean Sweeney (I knew Sean extremely well – he had been a corporal in my platoon in Aden).

PART TWO: HAVING GOT THE HANG OF IT

1985 started at the same quick pace as its predecessor, but it gave the new team an early chance to shake down. By 4 January the battalion was back at Soltau, for another major training period. Throughout this memoir the names Soltau and Sennelager crop up repeatedly. They were the main training areas used by the British and the difficult trick, after so many visits, was to find something different to do – repetition would take the edge off whatever training package had been designed. I relished the challenge.

This time I decided to start with an escape and evasion exercise. It would be tough on everyone involved because the training area was under deep snow and night time temperatures were down to minus 27 degrees centigrade. Avoiding hypothermia would be a significant consideration. The 'runners' or 'escapers' would be platoon commanders and their sergeants, plus one or two other key personalities. Making them the runners meant that senior corporals would have to command platoons, or act as sergeants, and that would be a major step up in responsibility. In turn, lance corporals, or even senior private soldiers, would have to command the sections vacated by those corporals. The exercise was recognised as different and demanding and it was easy to discern an air of anticipation throughout the battalion.

I briefed the runners in Minden and set them off for Soltau in a forty-seater bus. They thought that they were going to their drop-off points, but they were in for a surprise. The bus driver had orders to pull into a lay-by just short of Soltau and there an ambush would be sprung. The driver stopped and jumped out of his vehicle, leaving the runners trapped inside. They were 'in the bag' even before the exercise had got under way. They were hooded and taken to some buildings used to practise house clearing drills, which had no windows or doors. The cold inside those buildings was intense. Waiting for them were some specialist interrogators from the Territorial Army, who would make life miserable for the runners for the next thirty hours. The interrogation phase would test runners' mental and physical resistance to their limits. I was interested to see how, after only a few hours, they had somehow shrunk within themselves. When permitted to move, they did so slowly. The cold made them hunch and with heads covered by hoods, it was difficult to identify people who I knew really well.

The rest of the battalion deployed to Soltau's hutted camp and made ready to move to its start positions. I had another surprise in store. I removed all the senior officers from their normal appointments. The RSM would command the battalion; company sergeant majors would command companies; the

Regimental Quartermaster Sergeant would act as the Quartermaster and so on. The senior officers would act as umpires, be available to their subordinates to provide advice if it was needed, and be prepared to step in to ensure safety, or to prevent embarrassment to people operating outside their normal experience.

I listened to Sean Sweeney giving his orders for the battalion to execute a night move of some 50 kilometres across the training area and to establish a line of observation posts (OPs) in an attempt to catch the runners. Frankly, I could not have done better myself and the company commanders reported that their sergeant majors had performed just as admirably. I got up next morning to inspect the line of OPs. The night move, in appalling conditions, had gone extremely well. The OPs were properly established and I was delighted at just how well a difficult operation had been carried out by the 2nd XI.

Meanwhile, the interrogations were over. The runners were again loaded onto the bus and were told that they were being taken for further interrogation elsewhere. The bus was ambushed yet again, this time by 'partisans', who freed the runners, gave them some basic instructions, equipment and rations and set them off in groups of three or four. Three days and nights of survival and escaping confronted them, while the battalion adjusted its OP lines and patrolled the areas that they thought the runners would cross.

At some of the rendezvous points, partisan contacts were provided by six intrepid wives from the battalion, who, after the Families' Day, had asked if they could accompany their men on exercise. Despite the cold, they played their roles brilliantly and, on getting home, told more war stories than their husbands!

One of the groups got lost and failed to make a rendezvous. Hours and hours went by, darkness fell and there was still no sign of the missing runners, who were all competent and experienced. I was increasingly concerned that a serious accident could have taken place and, given the severity of the weather, there might even be fatalities. I had always been confident of designing training that would stretch my soldiers, widen their experience, build on their abilities and increase their self-confidence. Had I overdone it with this exercise? Had I miscalculated the risk and put my soldiers in jeopardy? It was not a common thing for me to do, but I prayed – fervently.

The runners had to cross the artillery impact area at Bergen Hohne and firing was due to commence just after first light. I would have to tell the Range Control authorities to delay firing and was not amused when my Operations Officer told me that he had failed to clear our use of the impact area, thinking that the runners would be well clear before the scheduled commencement of firing. At the very last moment, the missing runners turned up, unharmed and able to complete the exercise, which subsequently proved to be remarkably successful. Most runners avoided capture and those who had occupied appointments well beyond their normal roles, performed surprisingly well, which deservedly earned them the respect of their superiors. New found respect worked both ways, because those who had 'stepped up' now appreciated just how difficult their bosses' jobs were. It had been a win-win exercise.

Variation in training was helped enormously by a wide range of adventurous pursuits. Foremost amongst these was skiing. Almost all units stationed in West Germany hired a ski hut in Bavaria for the winter months. The huts were basic ski lodges, often converted barns. Most were able to accommodate about thirty soldiers and their instructors. We sent over 200 soldiers on fourteen-day courses in both the winters that I was in command and all for the princely sum of £25 per soldier. It was no great surprise that *Exercise Snow Queen* was extremely popular.

So too were lesser adventurous training courses and one-off opportunities. Towards the end of 1984 a small group of soldiers slogged their way through the Alpine Pass Route, in Switzerland. The route took them some 200 kilometres from the north east of the country, near the border with Lichtenstein, to Lake Geneva. That pioneering group was so successful that a larger group followed them in 1985. An even chillier example was provided by one of the subalterns, Lieutenant Richard Clements, who took part in an Antarctic expedition lasting some four months, paddling his sea-going kayak amongst curious whales and chinstrap penguins, before working with the British Antarctic Survey. We also snapped up an opportunity to send a platoon to undergo a three-week course at the French Army's commando school. The French had some unusual ideas. I am not sure what good is done by having the belly of a tank brush over your prone body, but it was probably character forming and the boys valued their well-earned commando badges at the end of their attachment.

Sport continued to be as successful as it was in 1984. Pompadour boxers distinguished themselves by becoming divisional champions and runners up in BAOR. Hockey players nearly replicated their success of the previous year, finishing as divisional champions and Infantry runners up. The swimmers built on their successes, sweeping all before them to reach the Army finals again, while a newly formed water polo team fought their way through to the BAOR semi-finals. The athletics team reached the BAOR finals and not to be outdone, our Light Aid Detachment produced winning tug-o-war teams in the REME championships. The footballers raised their game, most notably bringing home a trophy from an international competition at Enschede, in Holland, where forty-eight teams from seventeen countries took part. And there was a new face on our sporting scene – in its first year, the orienteering team won through divisional and BAOR rounds to qualify for the Army final. In the Army's sporting world, nobody wanted to be drawn against a 3 R ANGLIAN team.

The other battalion in Minden was 2nd Battalion, The Royal Green Jackets, the direct descendant of my father's regiment, the King's Royal Rifle Corps. Ever since the Pompadours had arrived in West Germany, we had bettered them in anything that had brought the two units together. On the Garrison Sports Day, the Pompadours again won just about everything and it was abundantly clear, which battalion had the upper hand. It probably does me no credit, but I could not help thinking that if they had only known, perhaps those stuffy, senior officers, who had interviewed me for the Green Jackets when I was an officer cadet, would now appreciate that it was not at all wise to spurn someone simply because his father was a warrant officer. My first sergeant, Sergeant Brian Hutchinson, had a coterie of bon mots. One of them was 'It's a good thing to forgive, but remember the bugger's name'. I had done just that.

1985 was a very special year for The Royal Anglian Regiment. It marked the 300th anniversary of the founding of our most senior forebearer regiments. Actually, it was not the tercentenary for the Pompadours – our senior forebearer regiment formed in 1688 – nevertheless I decided to mark the anniversary in style. We would Troop our Colour and party.

Drill is the domain of the warrant officers and SNCOs. I told Mr Sweeney, the RSM, of my intention and he immediately recognised a challenge when he saw one. Drill had been bottom of the training pile during the APC conversion training in 1983 and in the helter skelter first year of armoured exercises in Germany. He would have to improve drill from a low base – a mountain for him to climb. Next, I appointed Major Rowland Thompson, 'C' Company's commander, to organise the celebratory events, then I took the family on leave to Spain and hired an apartment to have a wonderful seaside holiday at Sa Tuna, a picturesque village close to the French border. While things were getting underway in the battalion, it was better that I was not around.

On my return, things had moved on apace. The drill was hugely improved. Rowland's plans for an all ranks dance (held well before the parade), post parade lunches, and officers' and sergeants' mess balls in the evening, were imaginative and well advanced. I had invited Mr Norman Fowler, now Lord Fowler and Speaker of the House of Lords, to be the saluting officer for the Trooping – he had been a national service officer in the Essex Regiment – and he had accepted the invitation. The Colonel of the Regiment, General Sir Tim Creasey, and his wife were also to attend. Additionally, we had invited anybody who was anybody in local British military and German society to watch the parade. The pressure was on to pull this one off.

We were blessed with a beautiful summer's day. The barracks had been scrupulously cleaned. The parade square was lined with chestnut trees in full leaf and parked beneath them were armoured vehicles to 'hold the ground'. Stands had been erected for the hundreds of spectators and even when empty, the backdrop looked impressive. Those taking part in the parade had early morning physical training and a mandatory hearty breakfast as measures against fainting. Then the day began to unfurl properly.

The guests and our families arrived in droves. The battalion formed up, out of sight, behind the gymnasium and garages. I took over the parade, in its centre, from the second-in-command, turned the battalion to its right, took a deep breath and set them off. The Band and Drums struck up and I lengthened my stride in order to lead the parade as it came into view. The RSM marches second in the parade, six paces in rear of the commanding officer.

As I passed Sean Sweeney I said, *sotto voce*, 'What on earth are you doing here as RSM, Sweeney?'

'Look who's fucking well talking!' was his prompt and justified reply.

Two friends were enjoying themselves.

The parade went well. I remembered the many words of command, helped enormously by a soldier standing on a fire escape behind the stands, out of sight of the guests. He replicated what the soldiers on parade were doing behind me, so I always knew what point of the parade had been reached and what the next order should be. Having marched off and been dismissed, the soldiers joined their friends and their families for a barbecue lunch – the officers and sergeants messes had formal lunches with their guests. The officers' lunch was held in a large, ancient, khaki marquee. One of the officer's wives, Rosie Tansley, was a gifted, professional florist. She had insisted on a tent with large central poles that she could decorate; those decorations and the table arrangements were exquisite. If the noise of animated conversation was anything to go by, the guests had been impressed by the parade and by the lunch. As hosts, we were happy that things had gone well – so far.

The officers' ball in the evening was a family affair. We had hired a schloss on the banks of the River Weser, a few miles from Minden. We had also hired a river cruiser to get us there. The party made quite a sight as they boarded the boat, ladies in their ball gowns and men in mess kit. The battalion's band was able to form several musical groups and the jazz band played as the cruiser nosed its way along the river, carrying a cargo of increasingly happy people. As the party disembarked and entered the schloss, the band's fanfare trumpeters announced our arrival from the ramparts. It was a fairy-tale setting. We had a party to remember.

Celebrations over, it was time for the annual season of big exercises to begin. In September the tri-partite battlegroup again assembled at Sennelager for Exercise Treaty, to practice the contingency plan to advance along a closed autobahn to Berlin. The liaison visits between the national components and the study day had been completed and because I was now familiar with the way the exercise unfolded, I was confident of repeating the success of the previous year. There was one minor problem. Julian Lacey had held the fort for me while I had taken my leave and still had his to take. Later exercises would be more demanding, so I was happy to do without my second-in-command for the tri-partite training. I appointed Captain David Clements, my first adjutant, to act as second-in-command. He would benefit from the experience.

The night before the battlegroup left Sennelager to rehearse the convoy phase on an arterial route, I had given my orders and began a round of visits to the various components in their bivouac areas. The battlegroup was tactical i.e. soldiers wore their webbing and helmets and carried their weapons, vehicles were under camouflage netting and sentries were posted. I had made clear to everyone that during the exercise the battlegroup was 'dry' – no alcohol was allowed, except for the French who enjoyed a small amount as an integral part of their rations. I reached one of the French artillery batteries, accompanied by the French liaison officer. We immediately became aware that all was not well – no sentries challenged us, light escaped from many vehicles and tents, and there was a lot of noise. I stood at the tailgate of one of the larger trucks and looked in – a party was in progress. I told the liaison officer to find the battery commander so that he could explain the state of his battery to me. To my surprise, a man separated himself from the group inside the truck, staggered towards the tailgate and fell out almost on top of me. He was the battery commander and was barely able to stand. Through the liaison officer I made my severe dissatisfaction known and had my displeasure conveyed to the commanding officer of the artillery regiment, who was present in Sennelager to observe the exercise.

A short while later I was called for by the British brigadier who was responsible for organising the exercise. I was astounded to be upbraided for upsetting the French. Despite our difference in rank, I was having none of that. I described exactly what had happened and said that I was confident of his support. It was up to the French how they disciplined their errant officer, but I had the right to expect that the orders that I had given would be observed, whatever nationality was involved. The brigadier was less than happy, but my argument was difficult to refute. No more was said, but our relationship was terminally fractured.

The exercise progressed without further incident, but on learning that I had allowed my second-in-command to take leave, the brigadier accused me of not appreciating the importance of the exercise. I pointed out that David Clements had fulfilled the trust that I had placed in him and had performed admirably. That cut little ice and the brigadier said he would comment adversely in his exercise report. So be it. The exercise had been successful.

Almost as soon as the battalion had regrouped in Minden, we were off for another two weeks training in Soltau, as the precursor to the formation exercises. Again I tried to vary the training as much as possible. It had air mobile phases, using RAF support helicopters, and, as its highlight, a complex night river crossing was executed by the entire battlegroup. If they could undertake that successfully, they could cope with most things.

Over the second weekend the battlegroup moved, by rail and road,

150 kms south to a hilly, forested area close to the one we were earmarked to defend in war. We were joined there by 3 RTR and two Danish regiments, 2nd Zeeland Life Regiment and the 1st Guard Hussars. Together we exercised under headquarters 11th Armoured Brigade. Both the brigade and divisional exercises were fast moving and covered considerable distances, requiring frequent replenishment (replen) of combat gas – petrol and diesel – a difficult task, normally done tactically, at night and without lights. A single company would regularly suck up 3,500 litres of fuel daily. The administrative personnel in the battlegroup were severely tested to keep us moving, fed, watered and resupplied with ammunition.

Our train of resupply vehicles was organised and commanded by a woman officer, Lieutenant Jane Weir, who was the first woman to serve in the battalion, other than as the medical officer, in our almost 300 years. With the exception of women doctors and nurses, until this time women served only in the WRAC, but now they were finding wider roles across the Army. I had been asked if I would accept a woman officer in the battalion in a mainstream role. I did so provided that she could contribute fully, both for her sake and for ours. She was not big or strong enough to be an infantry soldier, so other roles had to be found for her. In barracks Jane became the assistant adjutant, but her flair for administration and her competence at map reading meant that she was a natural to run the replen. She gained professional satisfaction and our respect from doing the job well.

At the end of the divisional exercise, the battalion had been in the field continuously from 27 September until 1 November. Laundry was our first priority! And there was a bonus for soldiers living in barracks, in the form of a £164 refund for food and accommodation charges.

The return to barracks was welcome and it gave an opportunity to sort ourselves, weapons, equipment and vehicles out before they were subjected to a flurry of inspections by specialists from outside the battalion. Throughout my time in command, to ensure that all the battalion's kit was constantly kept up to the mark, I held company inspections. I was assisted by the quartermasters, REME SNCO mechanics and armourers, the chief clerk – who examined the maintenance of documents, and the RSM – who looked at anything that took his fancy. The team concentrated on finding as much that was right, as they did to uncover deficiencies. The inspections were not witch hunts, but served to reinforce good practice and eliminate weaknesses. They were serious, but could also have some lighthearted moments.

Shortly before Christmas I inspected 'C' Company. They were said to have a snake as a mascot, but it was kept well hidden. They were got out of bed early on the morning of the inspection and after various tests that they could have foreseen, I threw them a curve ball. I ordered them to Troop their snake, at a parade that they would design. In the afternoon the company would parade in their best uniforms, the drill would have to be good and they would be supported by the Band. The Bandmaster was aghast when told he had just a few hours to write and rehearse appropriate 'snake music'.

At the appointed hour, I left battalion headquarters with the RSM, to take my place on a saluting dais to watch the snake trooping. The company and Band were on parade, but there was no sign of the snake. I was received with a salute and then the Band struck up a tremendous wailing noise. From out of the company's accommodation came the snake, its handler and escorts, dressed as snake charmers. They had invented an extraordinary slithering gait to cross the parade square to their position immediately behind the company commander. 'C' Company were performing brilliantly. I looked around the square. It was lined with soldiers and more hung out of virtually every window. Everyone was laughing. The company marched past and the snake was extended in salute as it passed me. It was winter and the warm blooded thing was cold, so it barely moved, but it was returned to wherever it lived with no adverse effects. That episode may appear ridiculous, but it was fun and fun is like yeast – it lifts things – a vital ingredient for any demanding profession.

1985 was to be the last year that I served at regimental duty. I was blessed to have a personal and professional highlight before I left. Just weeks before I handed over the battalion, I was to command a battlegroup to train in Canada. From late spring until early autumn, a total of six British battlegroups, each of some 1,200 soldiers, were flown to Suffield, in Alberta, from their bases in West Germany, to conduct live firing exercises that in scale and realism, were unsurpassed by any other army. Each battlegroup was in Canada for about a month. Tanks, helicopters, APCs and other vehicles were pre-positioned there and it required the vastness of the Canadian prairie and the 2,700 square kilometres of range it provided, to accommodate the fast moving, wide-ranging exercises that the British undertook.

Preparations began early in the year. I was allocated two tank squadrons from 4th/7th Royal Dragoon Guards. 'A' and 'C' Companies came from the Pompadours, plus the Recce, Mortar, Anti-tank, Assault Pioneer and Machine Gun Platoons. To complete the line-up 4 (Sphinx) Battery would provide our close artillery support; a troop of engineers came from 29 Field Squadron; a troop of armoured engineers, with their specialist tanks, came from 31 Armoured Engineer Squadron; and two Gazelle reconnaissance helicopter crews were provided by 669 Squadron, Army Air Corps. All the sub-units brought their own administrative elements. I have listed them to illustrate just what a composite group it was. My first job was to weld this bunch of strangers, who had first met in January, into a highly efficient team by May.

The work-up training began with a TEWT (tactical exercise without troops), which involves commanders studying a tactical scenario, not in a classroom but on the ground, and then coming up with a plan. They have with them all the subordinate commanders and specialists they would normally expect for whatever operation was being considered. The TEWT had the advantages of getting my new team to know each other and of working together without troops and weaponry being involved. Unfortunately, I chose one of the coldest days of the year for it. While I was training my commanders, I also set a test for my cooks and mess staffs. By lunchtime those involved in the TEWT were frozen, but four marquees had been erected in the yard of the local fire station – one for the junior ranks, a second became the sergeants' mess, the third was the officers' mess and the fourth was the central kitchen. Everyone got an excellent four-course meal served in some comfort – well done the cooks and mess staffs. Our attached personnel thought they had struck lucky with the Pompadours and that was exactly what I wanted them to think.

Weeks later that excellent start was reinforced by exercises conducted at Soltau and mid-way through the fortnight there I was happy with the progress being made. The boys had worked hard, so the weekend break in training was kicked off with a battlegroup barbecue, brilliantly organised by the Quartermaster, Stan Bullock, and his chefs. The soldiers were allowed a couple of beers and it was good to see tank men, infantry soldiers, gunners and engineers playing impromptu sports and helping each other with maintenance. Team building was well underway.

The advance party flew to Canada in mid-May, but then a problem arose.

The RAF had to divert its airliners to another task, so there was a fortnight to wait until the main body's first flight would arrive. R&R leave was normally taken at the end of exercises, but once the advance party had completed its recces and the takeover of vehicles and equipment, they made use of the spare time by taking a few days local leave. I flew to Vancouver to come back through the Rockies by train, which dropped me off in the middle of prairie, where my staff car was waiting for me! That was a level of service I was never to enjoy again.

There were still some days to wait before the main body arrived, so Julian Lacey and I decided to combine a recce with a picnic by the Saskatchewan River. The place we wanted to get to was miles away from Suffield, so we tasked a *Gazelle* helicopter, with our picnic and an inflatable boat slung under the aircraft in a cargo net. Once the recce was completed it was time to set down the net and its contents. Over the intercom I heard the young officer pilot say, 'I've lost control on the tail plane.' His observer, a sergeant, verified what he had heard and asked the extent of the problem. I quickly gathered that we could only fly in straight lines and then only with difficulty. Turning had to be completed in the widest of arcs and we could not hover. 'I'll have to drop the net, sir,' the pilot told me. I thought that was the least of our worries, told him to go ahead and asked what was next. The two air crew conferred and together decided that we would have to make a 'run on' landing, like a conventional aircraft. They informed their base of our predicament. The helicopter had skids, but the prairie was rough and covered in tough scrub. If a skid caught in a big bush then the helicopter was likely to tip and should a rotor blade strike the ground that would almost certainly be curtains. We lined up on the flattest piece of ground that we could see and ran on. We bumped and bounced as the helicopter lost speed on landing. Bushes flashed past us, but the aircraft remained upright and slowed to a halt. The pilot immediately cut the engine and applied a brake to the blades. We had made it, which brought deep intakes of breath and a lot of high fives!

While the pilot and observer waited for mechanics to arrive, Julian and I recovered the net. Everything was intact. With nothing else for us to do, we inflated the boat and set off to find a picnic spot. On the way I thought I saw rapids ahead of us, but the rocks seemed to be moving and then they – the rocks – passed us. We did not pass them. Julian and I had been privileged to see sturgeon running and the 'rocks' were the backs of the huge fish, breaking the surface of the water as they made their way up river. Later, I was pleased to learn that both the pilot and the observer had been recognised for their calmness and expertise in a difficult situation. The awards were entirely justified.

Once the main body arrived, the battlegroup deployed. Away from the camp at Suffield, a long line of huge American-made trucks drew alongside the armoured vehicles. It was the ammunition arriving and the vehicles and soldiers were 'bombed up'. I was told that during the course of the next five weeks, the battlegroup would expend about £8m in ammunition. The training was as close to war as it was possible to make it and you needed to be well trained just to go and survive there.

Of all the exercises, I remember two in particular. The first was a battlegroup assault on a defensive position that replicated one that might have been constructed by a Soviet motor rifle battalion. There were myriad targets to eliminate with tank and artillery fire before the infantry could break in to the position and fight through a maze of trenches – exciting, even frightening stuff. The other exercise was my last at regimental duty commanding troops. It was a long armoured advance requiring pockets of enemy to be destroyed along the way. The companies and squadrons scythed through them before we made a concluding battlegroup assault. Realising that this was probably the final time I would have a rifle in my hands during a live firing exercise, I dismounted from my APC and together with Sean Sweeney, joined a platoon for the concluding attack.

Just before the battlegroup began its return to camp at Suffield, the officers and men were assembled for me to speak to them. At that moment we were, arguably, the most highly trained battlegroup anywhere in the world. I thanked them for their outstanding efforts and told them they should be proud of themselves because I most certainly was. To help me wind down from an incredible high, I changed places with my driver and drove my APC back to camp. It had been a wonderful day and a great way to make an exit from command in the field.

After the battalion returned to Minden, I had only days left as commanding officer. A round of farewells began. I was memorably dined out by the Corporals mess. They had no building of their own, so the meal was held in the soldiers' dining hall. Corporals represent the next generation of senior non-commissioned leaders and in saying 'goodbye', I wanted to make clear to them how much they had in prospect and what a remarkable, close and caring environment they lived in – they had a lot to gain and a lot to lose. I sensed that I had their close attention, but then, in a flash, it had gone. I was tapped on my shoulder and turned to have my head enveloped by two 38DD-sized boobs. It was difficult to know how loud the cheering was when your head was clamped in such novel and efficient ear defenders! The young lady in question turned out to be a student at Hannover University, supplementing her income. She was intelligent, pretty and a charmer.

I was also dined out by the warrant officers' and sergeants' mess, which was a difficult night for me. I had valued their support and advice since I was a subaltern. They were loyal to a fault and without them my time in command would have been far less successful and much less enjoyable. I tried, but could not find adequate words to thank them.

The final dining out was from the officers' mess. It was an occasion to savour. The dining room was as impressive as it had been in Ballykinler when I had first joined. The silver, which I now knew so well, glittered on the highly polished table, set off by a deep purple, velvet table-runner. The Colours, on their stand, provided a familiar backdrop. The ladies were in their finery and the men's scarlet mess jackets completed a scene of colourful splendour that would be difficult to replicate outside the military.

When the meal was over, it was time to listen to the Band's after dinner music. Diners gathered in a marquee outside the mess and when the Band's programme had come to an end, the Bandmaster offered me his baton. It is the custom for any officer leaving the battalion for the last time, to conduct the regimental march. While *Rule Britannia* is an excellent march, I wanted the chance for a little bit more. Unknown to all but the Band, I had rehearsed with them and had learned to wave my arms more or less in time with the last couple of minutes of Tchaikovsky's 1812 overture. The overture's grand finale boomed out – the crowd were surprised and I was in seventh heaven. The overture came to its magnificent end and after a pause, I raised the baton to begin *Rule Britannia*. That was the moment when it hit home that I was really leaving. I had witnessed this scene so many times before for the benefit of others and now it was my turn.

Back inside the mess there was one last surprise – my leaving present from my officers. It came in a big box, which the gathered crowd insisted I unwrap. When the top came off, I lifted out the most horrible coffee pot that had ever been made. I looked up expecting a roar of laughter, but everyone was serious and straight faced. I had no option but to thank them for their generosity and was about to embark on a short farewell speech, when it all became too much for someone, who could not suppress a snort of laughter. With that the room collapsed and I had been caught at my own game. I loved practical jokes and now I was the happy butt of one. My actual coffee pot is silver, with the battalion's crest and a message from my officers engraved on it. It is a prized possession.

On my last full day in command, I cleared my desk, took out my headed notepaper and replaced it with that of my successor. I looked out of my office windows into the barracks and allowed myself a few minutes to reflect on the past two and a half years. When I had taken over as commanding officer the battalion was extremely professional in everything they undertook, but I had wanted more than that. I hoped that I had added the confidence and positivity that only success can bring. Above all, I had wanted them to relish their soldiering and to value one another. I thought I had achieved a good deal of that. In command I had only to flick the reins to get an instant response. I had ridden a thoroughbred. On the few occasions when times were tough, I could pull the battalion around me like a well-worn overcoat to provide comfort and keep out the cold. I could not have asked for more and felt humbled. I still do.

The following morning, on 11 July 1986, the last rites were performed. I was to be succeeded by a great friend, Alan Behagg. We had served together with the battalion in Aldershot and in Paderborn and were next alongside each other in the MOD, albeit in different branches. After I had attended the Australian Staff College and commanded 'C' Company in the 2nd Battalion, Alan had done the same. He seemed destined to follow me.

I took coffee with my officers in the officers' mess and then it was time to be pulled out of barracks. As I made my way to the front door of the mess, Julian Lacey waved a piece of paper at me.

'Look what's just landed,' he said. 'What do you think we should do?'

'Ask Alan,' I replied.

'No, Colonel,' said Julian.

He had been a superb second-in-command and a worthy successor to Peter Dixon. I could not differentiate between them. Both were talented officers, close friends and confidants. Their advice had always been candid, equitable and valued. If Julian wanted a decision from me, there would be good reason.

'You know the background, he doesn't. You decide.'

I did and that was me done.

I rode in the turret of a *Scorpion* light tank of the Recce Platoon. Two long lines of officers and sergeants extended in front of the vehicle, pulling on the ropes. We made our way along a route lined with Pompadours. They gave me a rousing send off. As we approached battalion headquarters, I looked up to see Sue in what had been my office window, and waved. Then there was only the barrack gate in front of me. The driver controlled the vehicle until its nose was just beyond the line of the gate. I climbed out of the turret, down past the driver's hatch and dropped to the ground. I was out of barracks and no longer in command. I had had the time of my life.

As a postscript and an indulgence, I would like to quote from a report written on the battalion:

> Successful conversion to the mechanised role is a complicated process involving detailed planning, close supervision and hard work from all ranks. There can be little expectation of a battalion acquiring all the skills in their first year. But the results achieved by 3 R ANGLIAN have been outstanding. From the outset their collective enthusiasm is deeply impressive and their rate of progress unusually fast... In short their professional impact within the Brigade and indeed the Division, has been considerable.
>
> They are the best example of a close knit military family it has been my privilege to observe.

It says it all.

CHAPTER 15

PLANNING FOR WORLD WAR III

Bielefeld is only 30 miles from Minden, but the change in my lifestyle and working environment in the next two and a half years would be marked. Gone were the staff car, the orderly to look after my kit and domestic help in the house. Sue and I were back to earth with a bump. First, we had a month in a flat, which, while inconvenient, was no hardship because much of that time was spent on leave in Italy. We hired an apartment in Viareggio, a seaside resort just north of Pisa and within easy reach of Florence. Our boys were happy with the beach; Sue was delighted to have so much culture readily on hand; and I simply unwound, helped by time with my family, good food and even better Primitivo.

The Military Secretary's branch – the people who deal with officers' postings – had entered me for two jobs after command. The first was as the military assistant to the Commander-in-Chief BAOR, a four-star general. The second, and the job I got, was as head of the operations branch in Headquarters 1st British Corps. I was pleased, notwithstanding the interest inherent in the military assistant's role. I am not a natural 'bag carrier'.

In the previous chapter I mentioned that a corps is a huge formation. When fully mobilised 1st British Corps would have had about 120,000 troops under command. To give context to my new job, I should explain how the Corps fitted into the allied military structure in West Germany. It was part of the Northern Army Group (NORTHAG), whose task was to defend the north German plain. Its area of responsibility extended from the mouth of the River Elbe in the north, to the city of Kassel in the south. NORTHAG had four corps resident in West Germany. Furthest north was 1st Netherlands Corps, south of them was 1st German Corps, and then came the British, with the small 1st Belgian Corps on the right flank. If World War III were ever to happen, all four corps would deploy to confront the Soviet 3rd Shock Army and its East German allies who, throughout the Cold War, lay operationally ready just the other side of the inner German border. Each of the four allied corps had its own plan for war, within a framework provided by NORTHAG. My new job was to hold and update the British plan. In order to do that I would work directly to the Corps' Chief of Staff, a brigadier, and very frequently, directly with the Corps Commander – a lieutenant general.

The operations branch in Bielefeld had a suite of four rooms on the first floor of the headquarters building, immediately adjacent to the Commander's and Chief of Staff's offices. The rooms lay behind a massive, steel door – visitors needed to buzz for attention in order to get in and, to all intents and purposes, the operations branch worked in a large safe. The steel door opened onto the clerks' office, presided over by a warrant officer chief clerk. He supervised four junior NCO clerks who worked to support the branch's staff officers. An adjoining room was home to three of those officers, the most senior of whom was a GSO 2, a major, attached from the Canadian Army. Canada does not have a corps, so this was a much-valued post that allowed one of the very best Canadian officers to gain experience of working in such a large headquarters. A second GSO 2, from the Royal Signals, specialised in electronic warfare. Finally, there was a GSO 3, who supported both myself and the Canadian officer, often picking up detailed, nitty gritty stuff that his bosses were pleased to avoid. The GSO 3 was Captain Mike Beard, a Royal Anglian, who had ridden with me in the Berlin Military Tattoo in 1979. Beyond the staff officers' office was my own. Its principal decoration was a floor to ceiling map of the Corps' area of responsibility. On the far side of the clerks' office was a secure briefing room and that was the extent of my empire. Quite a change from a large barracks, hundreds of soldiers and clanking armoured vehicles.

Our new quarter was a mirror image of the one we had had in Minden. It was located high on a hill overlooking the town. Its back garden had an extremely steep slope, so much so that I had to attach a rope to the handle of my 'flymo', to stand on top of the slope and swing the machine from side to side in order to cut the grass. The house needed some decoration, but I was told that there was a long waiting list. A quick call to the battalion solved that. I bought the paint and two of the Pompadours' domestic pioneers (highly skilled tradesmen) appeared and within days the house had a fresh feel, with all its minor maintenance jobs also sorted out.

Because of our move, Sue had given up her librarians' role at the Prince

Rupert School, but rapidly gained another at King's School, a corresponding Service secondary school at Gutersloh, about 12 miles from Bielefeld. To satisfy her constant need for a car and my occasional one, for the first time we became a two-car family with the arrival of her tiny, white Daihatsu.

The officer I took over from, Bill Marchant-Smith, left rather like Henry Wilson had done on handing me his job at the MOD. Both were keen to get away and I quickly understood why Bill was so fast out of the blocks. There was only one corps in the British Army and it was consulted on anything to do with war fighting. From MOD branches and from Headquarters BAOR came endless papers requiring comment. There was also a need for close liaison with our neighbouring corps that necessitated constant exchanges. The operational traffic between us and the divisions we commanded was even heavier. During the latter years of the Cold War, there were three British divisions permanently stationed in West Germany (1st, 3rd and 4th Armoured Divisions). On mobilisation, 2nd Infantry Division would complete the Corps' order of battle from its bases in the UK. With that wide range of customers, the operations branch was a busy place to work.

My initial job was to get to grips with the war plan. I needed to know its finest detail. The plan was disseminated to divisions, then to brigades and, subsequently, down to units on a need-to-know basis. Changes to the plan were made as new weapons and equipment enhanced capabilities and/or new commanders had different ideas about how to fulfil their war fighting missions. The plan was a vibrant, living document.

Within weeks of taking over, I was sent to England, to Porton Down, to attend a Nuclear, Biological, Chemical (NBC) course where I got an unusual amount of personal attention from the instructors. That was because as the head of the operations branch, I was the Corps Commander's principal NBC adviser. Up to this point in my career I had practised individual NBC drills and had taken part in NBC exercises at unit level, where the main aim was always to survive – and that was difficult enough. As part of my new job I was to advise on the use of tactical nuclear weapons – quite a step change. I paid particularly close attention to what was being taught, but good though the instruction was, I was not confident that I was sufficiently well equipped for such an important role. I need not have worried because in the artillery branch at Corps Headquarters was a specialist nuclear cell, headed by a major, who had a great deal of experience of serving in regiments with a nuclear capability. He worked for me and provided the detailed technical input I required when putting a plan together.

Once every year nuclear release procedures were practised during a NATO wide command post exercise (CPX) – a CPX requires only headquarters elements to deploy tactically and then to react, as they would in reality, to pre-scripted information fed from levels above and below them. The exercise was called *Able Archer*. It was always set in a doomsday scenario, beginning with relationships between NATO countries and their Warsaw Pact equivalents spiralling downwards, leading to the outbreak of conflict and a Soviet-led invasion of the West. At the highest level, heads of governments were involved in the final decision-making aspects of the exercise and corps headquarters represented the lowest level at which a nuclear response could be executed after appropriate authorisation had been received. For *Able Archer*, the Corps' nuclear artillery assets and their gun and missile crews were deployed, but they left their nuclear munitions in their bunkers. They made manifest the purpose of the exercise. Those crews would receive the ultimate order to pull a firing lanyard, press a button, or simply return to barracks without firing their weapons of dreadful destruction.

The exercise scenarios were written in great detail. They were as realistic as it was possible to be and included the awful impact on the civilian population that a modern conventional war would bring, as well as its dire military implications for the opposing forces. For the purposes of the exercise, it was assumed that the NORTHAG corps had all deployed to their war fighting positions, but after several days fighting, the weight of Warsaw Pact forces had pushed them back and military defeat became increasingly inevitable. Activity in Corps headquarters was frenetic with the operations branch in the eye of the storm. The political choice rested between being overrun and deciding on a nuclear response. The conventional exchanges were portrayed as being bitterly fought and the Corps had to redeploy its forces and liaise closely with what might be left of the civilian administration. West German civil servants assisted with the exercise to advise on a range of matters, including the movement of fleeing civilians, the treatment and evacuation of casualties, the provision of food, water and vital utilities and the disposal of the dead. Many of the civil servants had not contemplated the likely outcomes of World War III before their involvement in the exercise and found the graphic scenarios, of even the conventional warfare stage, too much to bear and left early. For them, to have to consider, in detail, an insight into a doomsday prophecy for their country and the certainty of huge numbers of their fellow citizens being slaughtered, must have been appalling.

The NORTHAG corps sought to contain the advancing enemy in large pockets, which would present appropriate targets for tactical nuclear weapons. Having taken a pounding, the various corps would make requests for nuclear release as the only means left available to destroy the enemy's forces. Those requests went up the chain of command to NATO headquarters and to allied governments for consideration. Given that requests came from several sources, that priorities had to be established, and assessments made of the possible enemy response, it was small wonder that it took time for release decisions to be made.

At Corps headquarters everyone was aware of when the exercise was scheduled to end and the closer that time came, the more imminent was the likelihood of a release order arriving. In every *Able Archer* exercise that I was involved with, a tangible sense of foreboding would flood the headquarters during this waiting period. Although the situation was not for real, people sensed the enormity of what was being practised. The physical effect on them was pronounced. Most spoke quietly and looked grim.

I went to the Corps Commander's caravan near the end of my first *Able Archer* exercise. Lieutenant General Sir Brian Kenny looked up expectantly as I opened the vehicle's door. 'Nothing yet, sir,' I said. He responded that it was difficult to bear contemplation of the awfulness that even a tactical nuclear weapon would bring about. Release would authorise him to give the order to fire, if he thought it was still the right thing to do in the military circumstances that prevailed at that time. If the war had moved on, conditions could be significantly removed from those we had predicted when making a request. The final responsibility would rest with him. Far from being gung-ho warmongers, the corps commanders whom I served, were compassionate and resolute men, who thought deeply about the extraordinarily heavy responsibilities laid upon them. On that particular exercise, release was withheld.

Work that daily involved planning for World War III, was a forbidding task and one that needed respite. Thankfully, the social life in Bielefeld was an active and varied one. There was always a round of private dinner parties to attend and a fashion for fancy dress made some of them memorable. For the thespians, Theatre 39 – a former Army cinema – provided an extremely good am-dram outlet. Sue joined its ranks for an excellent production of *Oklahoma*, then a pantomime and she also sang in the church choir. I turned to sport and regularly played football for the Army Crusaders BAOR, an all officer team. We mainly played against local German teams and the contrast was quite marked. They were skilful. We were fit. And that led to several confrontations when crunching British tackles took ball and man out of play, but it only took a few beers at the end of a match, to restore the bruises and goodwill.

I was persuaded to try a new sport – gliding. There was a gliding club at RAF Gutersloh, which is in the middle of the north German plain and the surrounding area is as flat as a pancake – not good for thermals, so the flying had to be precise. The club had a considerable number of gliders of different types, good hangar space and the run of the airfield for most weekends. The gliders were launched by winch and it was exhilarating to feel almost catapulted into the air and then sense the power inherent in those engineless aircraft once in the sky. It is often supposed that gliding is almost silent and serene, but that is far from the truth. The fabric covering the body of gliders, is stretched as tight as a drum and any air turbulence beats against it. The bangs can be very loud – alarming for a beginner.

I joined the club with a colleague, Alan Bush, who was already a pilot. With his previous experience, he made rapid progress. I advanced at a steadier pace. I could soon fly accurately enough, but had difficulty finding the weak thermals which were typical at Gutersloh and that meant my flights were fairly short, rarely extending beyond twenty minutes. I persevered and after a flight, without warning my instructor got out of the rear seat, secured his safety straps and said, 'You're ready to go solo!' He closed the Perspex canopy, gave me a thumbs up and stepped away from the aircraft with a smile.

There were butterflies in my tummy, but there was nothing for it but to point my index finger – the signal for the winch to take up the slack on the cable. I watched the tow line tighten and the aircraft gave a slight jolt as it became taut. I made a V sign with my right hand and waved it up and down meaning that the winch should go 'all out'. The glider leapt forward. I watched the airspeed indicator and as it passed the critical point, I eased back on the control stick. The glider rotated. With its nose now up, I steepened the climb to reach 1,200 feet before levelling off, then dipped the nose momentarily to release the cable. Free of its constraint, the glider gave a joyful little jump and I was piloting solo – wow, what a feeling! I even found an elusive thermal and my first solo flight was one of my longer ones.

All too soon I had lost height and it was time to turn down wind to return to the airfield. I made a second turn onto the base leg, to track at right angles to the final approach. One final turn and I was lined up with the landing area and chose the precise point to touch down. Next came a shuddering as the aircraft flew through turbulence created by ground clutter. I half opened the flaps on the aircraft's wings, then opened them fully and the glider stabilised immediately. I remembered my instructor's advice 'When grass looks like grass, round out [level the aircraft] and land.' Grass did look like grass, but I was passing it extremely quickly. Concentrate, concentrate … then grateful relief, the aircraft landed smoothly and in the right place. My instructor and several others ran over to congratulate me and shake my hand. I made to get out of the cockpit, but my instructor said, 'That was a really good landing. Bet you can't do a second one like that. Off you go!' Canopy closed, finger pointed, V sign waggled and I was up in the clouds again – and he was right. My second attempt at landing was safe enough, but quite a bumpy affair.

I continued gliding for a little while longer, but never really got to grips with finding thermals and as club members had rightly to chip in with all the tasks involved in running the activity, gliding always took the better part of a day. For me, it was too much of a commitment for too little satisfaction. I had achieved an ambition and earned my glider pilot's wings. I was content to leave my flying at that.

Our boys were growing up quickly. Nick was now thirteen years old and his time at Haileybury Junior School ended with the summer term. We had spent a leave in England looking at senior boarding schools for him to attend. The Junior School's headmaster had recommended three – we visited Harrow and Haileybury Imperial Service College. Both were impressive, but Haileybury had a particular feel about it and we did not feel the need to visit the third.

Haileybury's houses had only forty-eight boys each and they lived in the longest barrack rooms I had ever seen. The junior boy had bed space No 1. The head of house was at the far end of the room in bed space No 48. On either side of the room, boys' bed spaces stretched into the distance. They comprised a bed, cupboard, chair and a low, dividing screen and were not unlike loose boxes for horses. Washing and toilet facilities led off from the centre. The layout had changed little since the early nineteenth century when Haileybury was the Honourable East India Company's school for the sons of its administrators and East India Army officers. The housemaster's house adjoined the boys' dormitories, making him immediately available.

In addition to all this 'luxury', on joining the junior boys shared a study room. Study facilities improved as they got more senior, until they were allocated their own study in the Upper Sixth. The sixth forms had their own centre – in reality it was the school's pub. Each house included some girls, who joined in the sixth form and had their own accommodation. The school's academic record was good without being exceptional, but when combined with an outstanding arts offer and an excellent sporting record, we strongly felt it was a good 'all-rounder' and the school for our boys. Nick joined Trevelyan House in September 1986.

One of the most important tasks of my new job was to provide the crux of the Corps briefing to VIP visitors. There was a constant flow of them, necessitating two or three briefings most months, almost exclusively to politicians and senior British and allied officers. The briefing started with an intelligence brief, delivered by an Intelligence Corps lieutenant colonel whose aim was to impress upon the visitors the might and perceived readiness of the huge Warsaw Pact armies that were our potential aggressors. That done he handed over to me to describe the operational plan that we had developed to defend our part of northern Germany. The contrast in the opposing force levels was marked. Until reinforced, we had only three armoured divisions to provide a defence across a sector that was tens of kilometres wide and even greater in depth. If ever a war was to be fought in Europe, western governments would have to make an early decision on mobilisation and this was particularly true for the UK, because the deployment of reinforcements across the Channel imposed significant delay. The politicians hated the idea of having to make a speedy judgement, but the argument I advanced was virtually impossible for them to refute – delay risked a rapid conventional defeat and would place a greater reliance on the nuclear option. And there was more bad news for them. When I had finished, my opposite number in the logistic branch took up the cudgels to question the adequacy of our reserve stocks, particularly of ammunition and spare parts. He advised that to address those weaknesses would require the expenditure of many millions of pounds, at which point our visiting political masters frequently sank further and further into their seats. The briefing concluded with a civil affairs brief and while that was the primary concern of the host nation, it was yet more bad news. The unmistakeable salutary message was that given reinforcement, the Corps could hold out for some reasonable time, but without it we would be fighting a lost cause from the outset.

There was a light at the end of that dark tunnel in the form of 3rd US Corps. It has not been mentioned until now – it was the fifth corps in NORTHAG's order of battle and the most powerful army formation anywhere in the world. Just as 2nd Infantry Division was tasked to reinforce 1st British Corps, 3rd US Corps was earmarked to reinforce NORTHAG. The problem was that it was stationed in … Texas! But the Americans had a solution – they replicated an entire corps worth of vehicles, equipment and ammunition and stockpiled it in vast storage sites in Holland. That meant there was no logistic tail to follow 3rd Corps across the Atlantic. No other country could contemplate such an expensive solution. Its personnel had only to fly to Europe and pick up their kit, to be ready to fight, but 'only' represented a vast understatement – the task would have been enormous, requiring an even earlier deployment decision from the American government than that demanded of their European counterparts.

The 1986–87 winter was bleak and as the New Year began, Bielefeld was subject to the most amazing ice storm. I went to work as usual and mid-morning, I was working in my office when I heard a distressed groan, which was repeated several times. I walked through to the staff officer's room to find them looking out of the windows. I asked 'Who's making that noise?' Remarkably, it came from a large silver birch tree that grew immediately outside our offices. The tree was completely encased in ice and very fine, frozen rain was falling. The ice grew thicker as we looked out and the poor tree's groans got even louder and more agonised. Eventually, a large branch broke and fell to the ground. The barracks had become an ice rink and, even with care, it was impossible to walk more than a few steps without falling over. The icy conditions were such that our part of West Germany almost came to a halt, a position made worse because the federal authorities had decided that it was not environmentally friendly to use salt on roads, so the ice prevailed and, sadly, the decision caused many fatal road accidents.

Later in the year, Lieutenant General Sir Brian Kenny was succeeded as Corps commander by Lieutenant General Sir Peter Inge. General Peter was an outstanding soldier, but many regarded him as overbearing. I believe that he began his career as a private soldier in the Green Howards (an infantry regiment) and propelled his way through the ranks to be in the forefront of his contemporaries. His moods could be mercurial, but his military acumen and advice were exceptional. He took a very close interest in the deployment plan, so I saw much more of him than I did his urbane predecessor. Working with him was a bit like walking over quicksand – one was never sure of what was safe ground. At about the same time the commander of 3rd US Corps changed and Lieutenant General 'Butch' Saint arrived in Fort Hood, Texas. When Butch Saint met Peter Inge, they immediately recognised each other as kindred spirits.

I presume Butch Saint received a high-level briefing at NORTHAG, before he paid visits to the Dutch, German, British and Belgian Corps. He and his entourage were given the red carpet treatment when they arrived at Bielefeld and an already practised briefing had to be rehearsed several times before General Peter was satisfied. The American general heard the intelligence brief in silence and then it was my turn. I went through my spiel carefully, making clear our dependence on an early decision to mobilise and pointed out several advantages conferred by the terrain – wide rivers flowing directly across the potential enemy's line of advance and, in the south of our sector, areas of mountainous woodland, which would be difficult for advancing armour to negotiate. General Butch asked a couple of questions, which I answered satisfactorily. Then he turned to General Peter and said, 'My people need to know about this exactly as it's being told. Can you bring your team to the States?' That was readily agreed and planning began for a stateside visit later in the year.

Travel was the theme for 1987. In the spring, Sue and I returned to England to visit Haileybury Junior School to see Tim, soon to be Head of Choir, sing as Yum Yum, in *Madame Butterfly*. While boarding school provided stability and the friendship of contemporaries for our boys, we missed seeing many significant landmarks in their growing up. It made holidays all the more important when they came home. From their first years at boarding school, I was acutely aware of how much effort each of the four of us put into any time we had together. We were a close family and time has cemented that relationship and widened it to include our sons' wives, and their children. We are all certain of the love we hold for one another and feel privileged to be part of such a close knit, loving group of people.

I admit that I am hopelessly biased, but my sons have become the most impressive men I know; their wives have been easy to love as daughters; and our four grandchildren are a joy, developing as exceptional young people. Of all the many things Sue and I have accomplished together, our family is, by far, our greatest achievement.

I attended another NBC course later in the year, this time at a NATO school in Oberammergau. It had some value, but it did not compare favourably with the earlier course at Porton Down and pre-warned, I was able to take Sue with me, so it was as much a holiday as it was work. Our family holiday saw a return to Spain's Sa Tuna and by happy coincidence, Peter and Penny Dixon were holidaying at a property only a short distance away. Alan and Ros Behagg were with them and it was a memorable reunion. The six of us were already close friends and in the years since then, we have become almost as brothers and sisters.

The most remarkable trip of the year was made by Sue in October, in the company of some thirty officers' wives from Bielefeld. With the Cold War showing little sign of a thaw, they travelled to Russia. They flew with Aeroflot from Schoenefeld Airport in East Berlin, but their first stop was at British Headquarters in West Berlin for an intelligence briefing. Most of the wives were alarmed to be told that the Soviets would know exactly who they were, that their rooms would be bugged and they would be followed throughout their stay. Not all the ladies took the briefing seriously, but during their visit, suitcases were opened in hotel rooms and men in long coats made little effort to disguise the fact that they were 'tails'. The trip lasted about a week and began in, what was then, Leningrad, followed by an overnight train journey to Moscow, before several days of sightseeing in the capital. The visit gave a revealing insight into life in the Soviet Union. In contrast with the magnificence of palaces, churches and the Kremlin's architecture, were the paucity of goods, the grim, poorly maintained apartment blocks and the dilapidated vehicles. Sue thoroughly enjoyed the experience, but came home saying that she never again wanted to see anything covered in gold leaf! It became a standing joke that after the Russians saw our wives, they wanted to avoid the men and the collapse of the Soviet empire became inevitable.

In contrast to Russia, my big trip was to opulent America, to Fort Carson in Colorado. I went with a team of officers to provide input to a 3 US Corps CPX, which formed part of that Corps' work up training for a NORTHAG CPX the following year. We were taken aback at our American counterparts' over-confidence in their ability to destroy any Warsaw Pact threat once they had been committed to operations in Europe. They were acutely aware that theirs was the most powerful corps in the world and they thought that they could simply steamroller any opposition. They gave little thought to the ownership of real estate, which the four European corps would control, how to liaise and negotiate rights of passage and over flying permissions, of the constraints of European terrain (very different from the desert that they were used to), and how to move through defensive positions held by the European corps. To their credit, the American staff quickly appreciated that they needed to cooperate much more closely with us and the Dutch, German and Belgian Corps if they were to be successful. We needed their military muscle. They needed our local knowledge. Both sides recognised the deal to be done and good working relationships were established.

About six months into his time as Corps Commander, Peter Inge called me to his office. He told me that he intended to review the deployment plan personally. Only the Commander, the Chief of Staff, Brigadier Anthony Dennison-Smith, and I were to be involved. His review began with me briefing the plan in minute detail. I was encouraged to give my views and to argue my case – and it was not given to many people to argue with Peter Inge. There were three briefings in all, which resulted in amendments that I presented formally before they were adopted. The Chief of Staff then wrote to his opposite numbers in our four divisions instructing them to review their own deployment plans and that the Corps Commander would visit their headquarters in coming months, to hear the results of their appraisals.

When it came time for the divisional reviews to be presented, General Peter told me that only I would accompany him. That surprised me because I had assumed that Brigadier Anthony would be the accompanying officer. The reviews would reveal a lot about the thinking of the divisional commanders and their close staff and have a commensurately high profile.

Our first call was to 1st Armoured Division in Verden, commanded by Major General Rupert Smith, a charismatic Parachute Regiment officer. His Chief of Staff introduced the review, the detail was then covered by the principal operations officer before General Rupert summed up. I was sitting immediately behind General Peter. He did not turn to look at me, but simply said, 'What do you think of that, Colin?' General Rupert's head spun and he looked fixedly at me. From the reviews conducted at Corps headquarters, General Peter, Brigadier Anthony and I had identified the areas where we thought improvements could be made. 1st Division had covered them with one exception and had raised another that I thought would be well worth while exploring. I knew that General Peter knew all that and this was just as much a test of me as it was of General Rupert. I spoke my mind and General Peter concurred before adding some comments of his own. He was satisfied with the outcomes and the 1st Division team were rightly pleased.

In the helicopter on the way back to Bielefeld, General Peter said to me, 'You're a bloody good ops officer, Colin. By the time these reviews are over, you'll know how to command a corps.' That was praise indeed. During my time in Corps headquarters, I was to be on the wrong end of a couple of rockets from General Peter, an experience which felt like having paint stripper poured over you, but I knew that I had his confidence. Word had gone around by the time the other divisional reviews were held – everyone involved knew that although General Peter was the man to satisfy, I was the man to be watched!

In 1988, the level of unemployment was coming down in the UK, but remained above 2.5 million; the NHS was in its perpetual financial difficulty, with its nurses taking industrial action; and the IRA's campaign continued unabated, both on the mainland and in Northern Ireland. One bright spot was the house price boom in London and the South East, with a year-on-year increase of almost seventeen per cent. Our house in Colchester had always had tenants and had provided a good return. Its value had increased nearly fivefold and Sue and I thought that the time had come to take a further step up the housing ladder. We put our house on the market and quickly agreed a sale. We then took a short leave in Colchester to find another, slightly larger property and found one marketed by the agent who had looked after our own house. He told me that we had engaged the worst solicitors in Colchester to sell our house and we really should find another, much more proficient firm, to help us buy the next one. He recommended a friend and telephoned him to arrange an immediate appointment and I went to see Roger Buston, whose office was crammed with Royal Signals paraphernalia – he was a Territorial Army officer. When he ascertained that I had commanded the county battalion, he could not have done more to help.

Our previous mortgage lender had refused to give us a mortgage to buy the second house on the grounds that we lived abroad. Theirs was a fatuous position – I was a UK taxpayer – nevertheless, it presented a problem. Soldiers move often and rarely have time to establish an 'old boy' network, but Roger introduced me to one that solved the mortgage glitch instantly. He rang a friend who managed the Colchester branch of a UK-wide insurance company. He explained the circumstances and put the phone down to tell me that I had an appointment with the manager the following morning. I had only to sign the mortgage papers – the mortgage had already been agreed and I had not spoken a word.

Our luck was not to end there. The next day, the manager had an agent, armed with a new-fangled, then barely mobile phone, outside the property that we wanted to buy. Once I had signed the mortgage papers, the agent was instructed to complete a quick survey (the house was only eighteen months old and still settling, so an in-depth survey was not required) and a sale could be agreed. The seller wanted to stay for another three months, while his new property was completed. With house prices rising by thousands of pounds monthly, that was bad news, but the insuranc ˜nager held him to the agreed sale price and – bingo – the house was our?

The professional highlight of my last y' ˙ranch was the NORTHAG CPX. It involved all f than most similar exercises, lasting a f because the exercise opened with the advanced; the European Corps were a armies; and 3 US Corps had deploy vehicles, weapons and equipment. Warsaw Pact aggression, it requ' to launch 3 US Corps and driv previously conducted at Fort (passed off successfully.

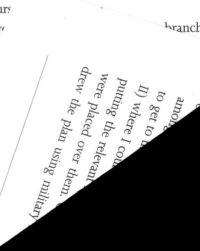

The British had one huge advantage over other corps anywhere in the world. In 1985 the *Ptarmigan* communications system was introduced into service. It was a battlefield wide, secure, mobile, digital, modular system that supported radio and data transmission. In the two to three years since its introduction, the Corps' staff had learned how to get the best out of that revolutionary communications system. It was a game changer.

Once the CPX battle was joined, every morning began with a corps update, delivered to the Commander and his senior staff. It began with a weather forecast from our resident expert attached from the Met Office. Even today adverse weather can so impact military operations, that it was imperative to get that briefing out of the way before considering anything else. An intelligence brief followed, then I would outline the operational situation and give a forecast of how fighting might develop in the next few hours. The logisticians would invariably tell a tale of woe about how difficult it was to sustain such intense levels of combat and then the divisions would be brought in to update their positions. While there was a physical meeting of officers based in Corps headquarters, the outliers would join using an early form of video conferencing, carried by *Ptarmigan*. It generally worked well. When the briefing was over the Commander would direct what he wanted to happen next and a planning team immediately gathered to translate the Commander's direction into operational orders.

The planning team consisted of an assistant chief of staff – a colonel – who acted as chairman. Working to him were sixteen lieutenant colonels, representing each of the arms and services e.g. intelligence, operations, armour, infantry, artillery, engineers, communications, logistics, medical etc. All had their own priorities and reconciling those was a delicate balancing act requiring everyone to accept sensible compromises quickly. Once the plan had been hammered out in reasonable detail, which took an hour at most – I took over and had about another hour to write the orders. I was *primus inter pares* ...gst the lieutenant colonels and could draw on their specialist expertise ...he detail. I had my own caravan (a bit like Monty's in World War ...ld work undisturbed. It had large map boards and I began by ...maps on the boards, then sheets of clear, pliable plastic ...Once the map co-ordinates had been established, I ...symbols to indicate the units and formations involved; arrows to show the movement required and routes to be taken; more symbols to denote positions to be occupied and so on, until I had a diagrammatic depiction of the operation. The operational diagrams I produced always needed expansion, so on areas of the map not covered in symbols and arrows, I wrote in vital details like the mission, concept, timings, who should move in what order and whether or not radio silence was to be imposed. A quick call to the Commander and/or the Chief of Staff to check my work and then the plastic sheets were numbered in the order they were to be transmitted and off they went to the signallers to be sent securely to the divisions and anyone else who needed them. As days progressed and the battles ebbed and flowed, the Commander would give further direction and the planning and orders process would be repeated. With the level of data transmission afforded by Ptarmigan, we were light years ahead of other corps. We had no need for the despatch riders, who they still had to employ to convey written and/or diagrammatic orders, although that alternative was still available to us should the technology fail.

When deployed, 1st British Corps had two identical headquarters because it was necessary to move every twenty-four hours or so, to avoid being located by the enemy – our visual, thermal and electronic signatures were extremely difficult to conceal. It was impossible to exercise command while on the move, so two headquarters were required, one to be in active command while the other set up, ready to provide a headquarters in a new location. If the headquarters could be located in buildings, it helped co ⁿⁱ its physical, electronic and heat signatures because the shielding insi ⁿⁱs so much more efficient than camouflage netting in oper Corps headquarters normally found itself in dis

A transfer of the headquarters to a n day of the NORTHAG CPX, when Ge advantage of having all of his most se that had nothing to do with the exe 'brass', the planning team's lieuter left in Corps headquarters. Some ranking people should accompa temporary commander. Genera' manage for the short time inv transfer of command to be made. I picked up a microphone to speak to the old headquarters and I said, 'I am ready to assume command.'

The response came, 'You have command.'

To confirm, I said, 'I have command,' and, for the one and only time, I meant it literally.

I was never close to being a lieutenant general, but for two or three hours, until the top-level meeting was over and the 'brass' caught up with us, I commanded a corps!

Sue and I continued to travel when we could. 1988 saw us take short breaks in Trier and Strasbourg; in Brussels to stay with Paul and Alison McMillen (Paul was now serving at Supreme Headquarters Allied Powers Europe); and then in Bruges. The big family holiday was in Baden Baden, where we hired a cabin on an activity site that pleased Nick and Tim and we were also able to take in some sightseeing, a concert and an almost mandatory visit to the famous spa.

Tim was now thirteen and after the summer holiday, he caught up with Nick and moved to Haileybury's senior school. As teenagers they felt that they had out-grown the mass of Lego they had collected over the years. They had looked after it carefully and now wanted to sell it. While the Lego was broken down and stored en masse, the boys had kept all the instructions. Both their grandmothers and Great Aunt Hilda were staying with us at that time and they were pressed into service to make up the various Lego designs. When each design was completed, its bricks were put into clear plastic bags, together with the instructions and a sale price was attached. I placed a short advertisement on a notice board at the entrance to Corps headquarters and pandemonium broke loose. The home phone rang constantly, as did my office phone. Buyers came to the house, but others elected to pick up their purchases from my office. The operations branch had never had so many callers. If the Soviets had invaded then, nobody would have noticed – Lego priority.

st important family success that year went to Sue. Soon after our en, she had enrolled on an Open University degree course. ll time job and having to meet the demands of being 's wife for much of the intervening time, Sue had e reward of a well-earned BA. Nick and Tim were quick to recognise their mother's achievement and made her 'Woman of the Year 1988'. Their homemade banner touched her and still does when the photo albums come out.

Theatre 39's end of year pantomime was *Snow White and the Seven Dwarfs*. As always the production was first class and Sue was part of the cast. But the show was stolen, not by talented, amateur actors, but by a major general and six brigadiers. They came onto the stage as the dwarfs, moving awkwardly on their knees and singing 'Hi, ho, hi, ho'. Their makeup and costumes were excellent and it was the best of best kept secrets. It brought the house down. These things happen all too rarely, but it is wonderful when genuinely important people elect not to take themselves too seriously. Well done them.

I had served two and a half years in my appointment. It was time to move on and excellent news – I was to be promoted to colonel. I was given a choice: another operational staff job, this time in a NATO headquarters in Izmir, Turkey, or as Commander, Brigade and Battlegroup Trainer BAOR, in nearby Sennelager.

Sue and I had been in West Germany for five years and a change of venue had real attraction, but I was fed up with shuffling paper and I liked being the boss. I knew the local military scene and Sue could retain her job if we stayed in Germany. Sennelager was our mutual choice and we made another short move in February 1989.

CHAPTER 16

VIRTUAL WARFARE – THE BRIGADE AND BATTLEGROUP TRAINER

We WERE ALLOCATED a quarter at 5 Arndt Strasse in Paderborn. It was exactly like the ones in Bielefeld and Minden, so setting up home was easy. Sue was always able to make even the worst of our quarters look special. Number 5's colour scheme did not suit her taste and fortunately we were able to convince the powers that be that some redecoration was needed. Once that was completed, the house had become a home.

While the houses were identical, the one in Paderborn had the best of the three gardens and it was probably there that my lasting interest in gardening was sparked. The statement items were two weeping willows – messy because of the amount of detritus they shed, but beautiful nevertheless. There were areas where flowersbeds could have been developed, but essentially it was an informal garden, so flowering shrubs and under planting with bulbs were better options than attempting herbaceous borders.

Arndt Strasse was about three miles from the training centre at Sennelager, where the Brigade and Battlegroup Trainer (BBGT) was situated. That posed no problem because the job came with a civilianised, green Mini. For Sue, her journey to work at Gutersloh was extended by about ten minutes. The 'Little NAAFI', a small supermarket, was conveniently only about one hundred yards from our house and Paderborn had grown and become more attractive since we were last stationed there in the 1970s. All in all, it was a pleasant place to live. We had made the right choice in electing not to go to Turkey.

One of the Army's principal strengths is the effort and resources it devotes to training. The two BBGTs – one in the UK, the other in West Germany – were early examples of the way developing technology had been embraced to enhance the training of commanders and staff. The training involved close scrutiny of the planning, decision making and control processes, both before and after battle had been joined. To achieve that, tactical scenarios were written that took place on ground within a few miles of the respective BBGTs. Commanders were required to recce the ground and make a plan in conjunction with their intelligence, operational and logistic staffs. That plan would be translated into formal written orders and subordinate commanders would be briefed. A battle would then be fought against simulated Warsaw Pact forces that would require those being exercised to respond rapidly to developing situations and to issue appropriate orders to counter, or take advantage of, changing battle conditions. All of that was carefully monitored by the BBGT's staff for the four days that the various headquarters were with us. During that time two battles would be fought; the second was a much more hurried affair than the first, requiring rapid planning and hasty execution.

Both trainers were quite modest establishments. The centrepiece of mine was a control room located in a very large hall. Occupying its middle, on a low, raised platform, was the map of the area over which a scenario would be fought. The map measured about 12m × 6m. The map sheets were laminated onto large, wooden tiles that in a matter of minutes, could be changed to reflect whatever ground was needed for the selected scenario. On all four sides of the map was a mezzanine level, partitioned into cells at which controllers sat. At the head of the room were the BBGT staff who would input information from the level above that being exercised. On the other three sides were subordinate commanders and specialist cells such as infantry, armour, artillery, engineers, aviation and logistics. They fed in information from below the exercised level. They observed and reported the battles as they developed in their particular areas, with symbols depicting the opposing units being manoeuvred on the map to reflect the ebb and flow of conflict. The symbols were frequently removed, or reduced, as casualties were awarded.

The exercised headquarters was located in another part of the trainer. To add realism, wooden, life-sized mock-ups of APCs had been built in former garages. They were under camouflage netting and laid out as a headquarters might be when deployed. Inside the mock-ups, it was difficult to tell them from the real thing. The officers and soldiers of the headquarters being exercised wore combat uniform, helmets and webbing. They had no access to the control room. A sophisticated, computer controlled, radio loom connected the exercised headquarters to its higher and lower controllers and the loom took note of the ground over which the exercise was being played. If, in reality, a hill got in the way of a particular signal, then that signal would not get through on the exercise. Another computer programme umpired the battles and forecast the results of the exchanges between the red (enemy) and blue (friendly) forces. Blue was never allowed to win. The BBGT staff would give credit for timely, correct decisions and clear, succinct orders, but they intervened to ensure that something always went wrong, just sufficient to keep the blue commander and staff on the back foot and working under pressure. The trick for the BBGT staff was to gauge just how much pressure the headquarters being exercised could take. The trainer had to provide a positive learning experience, or there was no point in the various headquarters coming.

The staff at BBGT were carefully selected. Most were majors who had commanded successfully at company/squadron level. They came from the infantry, armoured corps, artillery, engineers and logistic corps and they closely monitored the work of the corresponding staff in the exercised headquarters. Because they were recognised as being experienced and competent by the people being exercised, their advice and criticisms were readily accepted. There were also two warrant officers and a SNCO who were vital to the team. One was a hugely experienced warrant officer Class 1 from the Intelligence Corps. He was the 'red commander' and he manoeuvred the enemy forces during the battles. He also observed the work of the exercised headquarters intelligence staffs. The second warrant officer and a SNCO both came from the Royal Corps of Signals. They were the technical wizards who looked after the computers. Without their expertise the whole set-up would fall apart. In addition to those people, I had a major who looked after the day-to-day running of the trainer, liaised with the headquarters to be exercised and pulled together the first drafts of the formal reports that went to the exercised commanders within days of their visit. He supervised a small group of seven – clerks, administrators and cleaners. I liked the staff and concluded that I would have no problem making them my team.

During my time commanding the Pompadours, I had exercised at BBGT (BAOR) annually. That was standard practice for brigades and battlegroups – nearly forty headquarters were exercised every year. Frankly, I did not much look forward to my visits. However hard I and my team tried, we always seemed to come in for a minor avalanche of criticism. We felt that we were quite good, but that was never reflected in our reports from the BBGT team. After my first few weeks at Sennelager, I realised that every headquarters that I had observed had been remarkably proficient. I interrogated my staff and they agreed and that that had been the case before my arrival. I concluded that the reports to exercising headquarters needed a radical change of tone.

A major advantage of the BBGT system had always been that the reports went only to the exercising commanders and copies were not sent to their bosses i.e. mistakes could be made in private. I could build on that. I wanted to make commanders and their staffs look forward to visiting BBGT, but first I had to educate my staff to highlight the strengths of headquarters and state clearly what they had done well, in order that those strong points be nurtured – and there were many. Only then was it necessary to focus on weaknesses and suggest how they could be addressed. Initially the team were a little indignant. They correctly claimed that strengths were mentioned, but I was able to illustrate that the balance of our reports indicated that there was far more to get right than to keep right. We needed to reverse that order. Weaknesses would still draw constructive criticism, but my staff should see the glass 'half full' and not 'half empty'. When that was done, word quickly got around that things were changing at BBGT (BAOR) and that brigade and battlegroup headquarters' competencies were a good deal better than had hitherto been thought.

I had also noticed that very little of the good work done in preparation for the exercise scenarios was saved. I directed that from now on any good intelligence analyses, excellent sets of written orders, map overlays, or logistic plans should be retained. Next I had our briefing hall redecorated and bought some first class, commercial display boards. The examples of excellence were attractively displayed on those boards in the briefing hall. It was pleasing to note how long officers from exercising headquarters spent pouring over them and it became a matter of pride to have an example of work displayed at BBGT.

Introducing a change of emphasis at BBGT and sprucing up the appearance of the place had me fully engaged. I probably had an overabundance of thoughts for improvement to the point that my staff would sometimes groan 'Not another GGI' (Groves' Good Idea). Notwithstanding that several had merit and gained the wholehearted support of my team as they were implemented.

My first months at Sennelager flew by and summer was quickly upon us. Nick had taken his GCSE exams and was pleased with his results. That got the holiday period off to a good start. As soon as they arrived from school, we sent our boys on a sailing course at the Mohnesee, in the shadow of the dam destroyed in World War II by the Dambusters. After receiving their certificates of competence, it was time to look further afield and all four of us embarked on an epic holiday to America and Canada.

We flew to New York only to find that our hire car company would not answer its telephone and seemed to have no pick-up point at the airport. After two hours of trying to make contact, we finally spotted a mini-bus bearing the logo we wanted. We were driven to some semi-derelict area of the city to be given a huge American limo and then launched into a five-lane highway at the height of the rush hour. To help us find our way, we had been provided with an A4 sized map of the entire USA. 'Keep the sea on the right' was my navigator's sound advice. Amazingly, it worked.

Our first stop was with Sue's cousin, Adrian, his wife Nina and their three children. They lived in Hingham, a short distance outside Boston. That was a particularly convenient location because not only could we spend family time with them, but Boston was home to the re-enactment regiment, the 10th of Foot of America. Some of their members had helped me with the Berlin Tattoo act in 1979 and I was keen to renew our acquaintanceship. The combined efforts of Adrian's family and officers of the 10th of Foot ensured that we had a tremendous time. We were generously hosted and thoroughly entertained and gained insights into American history at the Plymouth Settlement, Boston Harbor (of tea party fame) and Lexington Green, where the first shot of the American Civil War was fired.

We left Boston to drive through New Hampshire and Vermont and on into Canada, to Ottawa, to stay with Roman and Irene Jacubow. Roman and Irene had been with us at the Australian Staff College in 1978 and we had maintained our friendship. They too looked after us royally and were eager to show us the sights that Ottawa had to offer, although the boys and I began to weaken when yet another museum, or gallery, appeared on the visiting list.

The return journey to the States took us via Niagara Falls – breathtaking – and then to Washington and our last visit of the holiday with Jonathan and Joy Hall-Tipping. Again, our hosts had put a lot of thought into our visit. Washington was impressive. Not only did we take in the White House, the Lincoln Memorial and the National Air and Space Museum, but we were treated to an open-air concert at Wolf Trap and a visit to Jonathan's office. His last posting in the Army had been on the military staff of the British Embassy in Washington. He left the Service and rapidly became a leading stockbroker on the New York stock exchange. He was keen to show me how he had done it. I was fascinated to see that he had simply applied systems to his new career that he knew worked. His principal administrator was 'the adjutant'; the person in charge of the teleprinter, computers and other communications was the 'signal officer'; the clerks' office was the 'orderly room'; and the brokers were equivalent to company commanders, with 'A' Company possibly dealing in futures and 'B' Company selling derivatives etc. Some might have thought his method quaint, but it worked – my, how it worked. He gave our boys a few dollars at the start of the day and they had made a small profit by its end. While they were impressed with that, Jonathan's swimming pool and his sit-on lawn mower were the high spots of their stay in Washington.

The end of our trans-Atlantic sojourn did not mean that holidays were over for Sue and me. In October, we spent a few days in Paris, staying with Tim and Liz Robertson, our next-door neighbours in Camberley, when Tim and I were fellow instructors at Sandhurst. Tim was now the assistant military attaché at the British Embassy and we could not have had better guides to that romantic city. From Paris we made the short trip to Mons, in Belgium, to catch up with Paul and Alison McMillen – keeping staunch friends close was, and is, vitally important to us.

Sue had spent nearly eighteen months as the librarian at King's School, during which time she had computerised the library's catalogue of 9,000 books. She enjoyed the job, but had become increasingly frustrated with her paltry remuneration – the MOD paid less for the qualified UK librarian than it did for the German cleaner – and she had resigned at the end of the summer term. She had a burgeoning interest in stocks and shares and finance generally. Shortly after our return from America, she secured a job with the Armed Forces Financial Advisory Services, a UK-owned, commercial company that focused entirely on the military market. After training, she was in her element, speaking at officers' and sergeants' messes and advising individual clients. She could not have been happier and our own finances took a turn for the better. Unsurprisingly, she established a professional affinity with Jonathan Hall-Tipping, which was to become a business partnership in later years.

I have previously mentioned that having our sons at boarding school meant that Sue and I missed many milestones in their growing up. We did all we could to attend the more significant ones and I will recount three in an attempt to demonstrate how important it was to all four of us that we were present.

The first was when Nick was about fourteen years old and Trevelyan House held a cross country run. Nick put in an enormous effort. He did not have the easy grace of the natural athletes in front of him, but less gifted runners were not going to pass him. As he came down the home stretch, somewhere in Hertfordshire the needles of seismic measuring devices quivered as he pounded the ground to force himself into fifth or sixth place. His housemaster appeared at my elbow and said, 'He's never done that before. That was for you!'

Next, at a Speech day, it was reassuring for us to talk to their housemaster and see both sons happy in the school environment. We visited the art block and were delighted that an entire display was devoted to Tim's work. We realised then that he had a real talent that was to underpin his later career as a graphic artist. His rather brutalist self-portrait, confidently and dramatically drawn at that time, hangs in front of me as I draft these words.

Finally, Trevelyan produced *Dr Faustus* as its house play. Haileybury is fortunate to have Alan Ayckbourn as an alumni and a theatre space at the school is named after him; a facility that any professional repertory company would be proud to possess. Sue and I struggled to make the 8pm start time and had to excuse ourselves as we belatedly took our seats. We had barely sat down when a spotlight shone on a young man on a scaffolding balcony, high above the stage. He lent on the scaffold railing and waited for the audience to settle. A hush descended, but there was still some shuffling of feet and the odd whisper. He waited until he had complete silence.

'*Had I as many souls as there be stars, I'd give them all for Mephistopheles!*'

A beautifully modulated, resonant voice delivered the opening line with a confidence that spoke volumes of the coaching that the young narrator had received. After a few seconds Sue whispered, 'I think its Nick!' It was Nick and it was the first time that we had heard his adult voice. Later in the play, a clown made an appearance. Tim had made his entrance and was wonderfully well cast, as comfortable in his performance as was his elder brother. We could not have been prouder.

1989 transpired to be a pivotal year in Europe. From mid-year onwards civil resistance in a succession of eastern European countries saw the Eastern Bloc implode. Rejection of eastern regimes began in Poland and Hungary, before spreading to East Germany, Bulgaria, Czechoslovakia and Romania, which had the only armed uprising. The Berlin Wall was breached on 9 November, heralding the formal dissolution of the Soviet Union in the following month. It was a remarkable time for anyone to be in Germany and that was particularly the case for soldiers, who saw long established, powerful, potential enemies disappear before their eyes.

By early 1990, East German cars were beginning to appear on West German autobahns in considerable numbers. Virtually all of them had notices in their windows with 'Freiheit' written on them. The smiling occupants of the all but cardboard Trabants and the barely better Wartburgs, gave thumbs up to their western cousins as they sped past in their Audis, Mercedes and BMWs. In the next few months it became clear that the two Germanys were destined to become one, as border controls had little significance and authority drained from the DDR's government.

I had not visited the inner German border since the breakthrough of the Berlin Wall. The opportunity came in July when Sue's brother, Roger, his wife, Helen and their two sons, Ben and Tom, came to stay with us. We set off for two villages, just south of the main crossing at Helmstedt, to view the border. I had chosen the villages because they were separated by only 300-400 metres, but the border ran between them.

The western village was prosperous. Its houses were well kept; its shops were colourful; its roads well maintained and everywhere expensive cars were evident. We walked beyond the village's eastern boundary to where a high, rigid, chain-link fence denoted the border. It ran north and south as far as the eye could see, but, directly to our front, panels of fencing had been removed. A rough road had been bulldozed through the fence and a deep ditch had been filled in. The road continued across a barren, wind-blown stretch of no-man's land to the eastern village. As my family carefully picked their way along the road's rocky surface, I could not help but reflect that there were no Volkspolizei, no Volksarmee and no Soviet soldiers to confront me. I was standing on their patch. Some twenty-six years after three young East Germans had made their escape through the wire in Berlin and I had decided to do all I could to confront the evil regime that they had left, we had won. It was an immensely satisfying moment – the professional highpoint of my career.

When we reached the eastern village, the contrast with its twin in the west was stark. A chain wire fence ran around the western side of the village, cutting it off even from no-man's land. Loud speakers were mounted on poles to broadcast whatever messages local party officials wanted to pass to the inhabitants; the roads were potholed; there was one petrol pump in the centre of the village square, but no garage. The pump was for the use of the Volksarmee border company and for farmers' tractors. I do not remember seeing a civilian car. Most telling of all – there was no colour. The doors and window frames of all the buildings were virtually devoid of paint. It had peeled off over decades, never to be replaced. At the edge of the village, the red ring that should have bordered the 50 kilometres an hour road sign, had faded into oblivion. A small sign, depicting a beer, was the only advertisement. Everywhere looked and felt drab. It was a sad, sad place.

While Germany looked forward to reunification and better times, Kuwait was to suffer the invasion of Saddam Hussein's Iraqi forces on 2 August 1990 and an occupation followed that would last seven months. The outcry against that illegal, brutishly aggressive act was such that a coalition of forces from thirty-five nations was brought together relatively quickly. The initial British contingent was 7th Armoured Brigade, based in Soltau. Individual reinforcements brought the brigade up to its war-time strength and work up training began in Germany before the brigade's vehicles and equipment were shipped to Saudi Arabia. Part of the early training was for the expanded brigade headquarters to exercise at BBGT.

My staff and I wrote a new exercise scenario for 7 Brigade. The ground could not replicate desert conditions, but the exercise helped the headquarters to shake down. With so many reinforcements drafted in, a significant number of the brigade's staff barely knew one another on their arrival at BBGT. The intensity of our exercise quickly brought them together and the brigade commander, Brigadier Patrick Cordingley, was happy with the progress his team made in the short time that they spent with us.

That was the sum of BBGT's contribution to the first Gulf War, but not so for Sennelager Training Centre. The British contribution to coalition forces was soon increased and 1st Armoured Division was committed to the operation to liberate Kuwait. As Sennelager provided the biggest and best infantry firing ranges available to the British in Germany, it was at the heart of the training effort for the much bigger force. The ranges banged away for long hours, seven days a week. While the sound of constant gunfire is redolent of that time, the all-pervasive smell of cooking bacon is even more evocative as the camp's kitchens provided a 'full English' twelve hours a day, every day, with lunch and dinner being served too. The place never slept.

With some notable exceptions the local German population was not particularly supportive. Their focus was much closer to home and on re-unification, which formally came about on 3 October.

In anticipation of that momentous act, for much of the year Soviet forces had been withdrawing from their bases in East Germany and returning to their homelands. This generated a more relaxed atmosphere and led to a visit to 1st British Corps by the commander of 3rd Soviet Shock Army. He wanted to observe some of our training and was brought to Sennelager. BBGT was on his itinerary.

Soviet generals wear the most enormous hats and under an example of such outsized headgear, a squat individual strode towards me as I stood outside the trainer ready to welcome him. Soviet uniforms were of poor quality and were badly tailored. The general's rumpled appearance was made worse by his tunic being weighed down by a vast number of medals and stars attached to it. He smiled a gold-toothed smile and shook my hand. 'Spitting Image' would have had a field day characterising him!

Be that as it may, he was every bit the soldier. He took a close interest in what we did and his visit overran significantly. He asked if I would be able to evaluate the efficiency of the headquarters of a Soviet motor rifle regiment. I said that given time to understand its procedures and enough interpreters, after a few days exercising, I could tell things about its staff that even their mothers did not know. He laughed and it was my turn to ask a question. I asked if he had revised his opinion of the British Army as a result of his visit. He gave an emphatic 'No' and added that although we were remarkably well trained, we were far too small. While we could fit all the commanding officers of our tank regiments into one mini-bus, we simply were not big enough to cause him concern. He was nothing if not straightforward and his remark was not lost on the senior British officers accompanying him.

Before the general left, he kindly presented me with a book on the Soviet Army and a desk tidy. The latter object had almost certainly been made in a

Soviet military workshop and was wonderful because it was so appallingly dreadful. Something resembling a red plastic tray had been used as its base; copper wire had been bent to hold items of stationery; and even the screws, which held the contraption together, were mismatched. It was an absolutely unique object and it was given pride of place on the wall of the briefing room in BBGT.

As 1990 drew to a close and 1st Armoured Division made its way to the Gulf, life at Sennelager slowed. BBGT continued its evaluation of the brigade and battlegroup headquarters that remained in Germany and my expectation was that this normality would continue into 1991. But while that was to be the case for BBGT, it was not the case for me. The commandant at Sennelager Training Centre was Brigadier Dick Mundell and he was due to be posted in January. My two best Christmas presents were to be promoted to brigadier and to become Dick Mundell's successor.

CHAPTER SEVENTEEN
TWINKLE, TWINKLE LITTLE STAR

THE VARIOUS RANKS of general are denoted by 'stars'. A general is a four star; a lieutenant general is a three star; a major general a two star; and a brigadier a one star. I was now a brigadier, a one star and a general, albeit a junior one.

I inherited two jobs from Dick Mundell. I became both the Commandant, Sennelager Training Centre (STC) and the Brigadier Infantry (BAOR). The appointments were interesting for different reasons.

STC was controlled and administrated by the British, who were also its principal user. It regularly hosted units from the Canadian, Dutch and German Armies. Its ranges were much better than adequate, but some needed updating. Its barrack accommodation was tired from years of over use and well below par for the late twentieth century. Improving both would be a challenge.

As the Brigadier Infantry I was the professional head of the Infantry in Germany. I had oversight of sixteen battalions and could visit them as I pleased. I was available to the Corps, divisional and brigade commanders to advise on infantry matters and, as a former and successful commanding officer, I also provided advice to the current batch of COs. As an adviser, I was outside the chain of command. That meant I did not write commanding officers' annual reports and whatever they chose to confide in me remained between us. With that assurance, most were quick to approach me and it was gratifying to help with their problems. That frequent liaison also meant that I remained in close touch with the views and needs of STC's primary users.

The move from Paderborn to Sennelager was the shortest of the four that Sue and I made in Germany. While the previous three quarters had been carbon copies of one another, Newmarket House was different. It was much bigger and more a residence than a house. It was discretely located in one corner of STC, away from the bustling activity of the main camp. The residence was quirky. I suspect that it was never intended as a quarter and it was much more likely that it was built as a small headquarters, or office block. All the bathrooms and toilets were confined to one corner of the building – a strange design for a residential building. It had two main floors and no cellar, again unusual for a

German house. The reception rooms, kitchen and bedrooms were generously proportioned. A fifth bedroom (Tim's) had been added in its vast attic. That room was a jerry-built affair, with partition walls and questionable heating, but Tim enjoyed his own space. The boiler room was sunk about a metre below the level of the ground floor and its pipework looked to have been taken from a trans-Atlantic liner. Outside there was a garage with storage underneath, two large areas of lawn and an attractive wild area leading to a stream. The wife of one of my predecessors had made an excellent job of establishing a large herbaceous border that needed only a little TLC to resurrect it. When Sue had put her stamp on the house – which included a lot of stencilling in the kitchen and bathrooms – it was a great place to live.

While settling into Newmarket House was a priority, I was keen to explore my new empire. Apart from BBGT, there were five other significant training establishments at STC ranging from the Northern Ireland Training and Advisory Team (NITAT) – designed to assist units prepare for service in the Province – to an environmental health school, which enabled its students to advise their commanders on how to keep their soldiers healthy in barracks, but more particularly, in the field.

STC was also a centre for recreational facilities. A Joint Services Parachute School was based at a former Luftwaffe grass airfield. The school ran free fall, adventurous training courses for Army and RAF personnel and for any dependants who were brave enough to jump out of an aircraft. Adjacent to the airfield was a first class eighteen-hole golf course, together with several football, rugby and cricket pitches. The BAOR stadium, where all the prestigious athletics and football finals were held, was also located within STC.

The live firing area is STC's *raison d'être* and lies immediately outside the environs of the camp. The training area was established by Kaiser Wilhelm II in 1892 and had been in continuous use by the military ever since. It is large, covering nearly 120 square kilometres. Because access to it has always been restricted and no intensive farming has ever taken place on it, particularly that involving the use of chemicals, the area has developed as one of the foremost nature reserves in north-west Europe. Helping to promote its ecology and to balance the needs of its ecosystems with those of the military became consuming interests for me.

I co-chaired a committee that oversaw conservation work on the area. The local German *regierung* (administrative district) provided the other co-chairman and around twenty ecologists – the area's bio-networks were so important that it's more sensitive parts were mapped, metre by metre, by those experts. That painstaking exploration revealed plants and animals in the live firing area that are not known to exist anywhere else in Europe, for example, translucent salamanders. Foremost amongst the specialists was the *forstmeister* (forest master), who was not only concerned with the management of large forested areas, but also oversaw hunting and fishing. The area boasted 2,000–3,000 head of deer, almost 1,000 wild boar and a few *mouflon* (large wild sheep), plus over 500 natural springs, several lakes and numerous streams – ideal for trout. The other environmental scientists included ornithologists, animal, insect and plant specialists, even a bat woman and a grass man. Grass man loved armoured vehicles. Their wide tracks make for very low ground pressure, but they grip and throw any loose plant material into the air. When seeds formed on grass heads in the autumn, armoured vehicles became grass man's ideal distribution agency. It was surprising how much we were able to help our ecologist colleagues, without detriment to our own activities.

A periphery road ran for almost a complete circle just inside the boundary of the training area. About twenty ranges of various types were immediately adjacent to the road and all directed fire towards the centre of the area. Some were purely for British use and were designed to hone marksmanship skills. The others all had tactical settings and required soldiers to move and fire as they would do in contact with an enemy. Soldiers of all the user nations exercised on them. At the lowest level were relatively small ranges, where individual soldiers fired their personal weapons and on others, practised with shoulder fired anti-tank weapons and threw grenades. Progression was made through larger ranges that exercised sections (8-10 men) and platoons (30 men), before the largest ranges could accommodate companies (100-120 men) with mortar, artillery and, sometimes, ground attack aircraft in support.

On 17 January, within days of my taking over as STC's commandant, hostilities began in the Gulf and were to last until 28 February. At Sennelager the training frenzy had reduced as 1st Armoured Division completed its preparations for war, but nevertheless it continued as battle casualty replacements were put through their paces. At the time, Saddam Hussein was supposed to possess chemical weapons and it was assumed that battle casualties might well be high. Individual soldiers, representing a whole host of disciplines, were earmarked as replacements and were trained accordingly. After speaking with many of them, it was clear they were unsettled, knowing that if they were deployed, it would be to fill the boots of others who had been killed or injured – not a happy thought.

Nationally there was enormous support for the troops and, as part of it, Sennelager was to receive a royal visit. On 31 January, Princess Diana visited troops and families in the nearby Paderborn garrison and a diversion was planned for her to call on troops under training at STC. I was asked to meet a royal recce party a few days before the visit, but the individuals concerned were agitated from the outset.

The front gate to STC is on the main road through the village of Sennelager. Directly opposite the gate is 'The Strip' – a long row of gaudy bars and neon-lighted nightclubs and strip joints, designed to entertain soldiers and rapidly relieve them of their money. The recce party deemed this row of dubious outlets to be sufficient to cause offence to Her Royal Highness. I was able to offer a solution, although I was far from convinced that it was necessary.

There was another gate to STC that was heavily overgrown and had not been opened in living memory, just short of where 'The Strip' began. It could be cleared, cleaned up and brought into service. The recce party were pleased, but insisted on seeing the old gate and the route that the royal party would take once through it. On the way to the gate we passed the military laundry, which was operating at full tilt, belching out clouds of steam. 'You'll have to shut that down,' I was told. 'Not a chance!' was my response. I explained that it was winter, wet and muddy and the boys needed their filthy combat clothing washed and turned around quickly for them to get on with their training. I was amazed that this explanation cut no ice. It probably did not help matters when I offered to have the steam coloured red, white and blue and the recce party and I parted disagreeing vehemently. Sadly, I was subsequently ordered to close the laundry for a few critical hours. I can understand that brightly illuminated depictions of boobs and bums on The Strip might just have caused the raising of a royal eyebrow, but I still cannot appreciate the objection to hot water vapour. I am certain Princess Diana was entirely unaware of the disruption her visit caused and would have been mortified to know.

Once Kuwait had been liberated and STC had returned to its normal schedule of training, I was able to investigate where improvements could be made to the ranges and barracks. Ranges were the priority. I was concerned that there were virtually no facilities to exercise the *Warrior* infantry fighting vehicle and *Scorpion* light tanks, with which recce platoons were equipped. Both mounted the 30mm *Rarden* cannon – a formidable weapon, but one that needed the skills of a tank's gunner to operate it, so the deficiency in STC's facilities was a major one.

I identified a suitable range for improvement, submitted my plans to the British authorities and to the local *regierung* and received their respective permissions to proceed. Work was well in hand when local conservationists began a campaign 'to stop British vandalism'. My design was for the construction of trackways along which vehicles would move and fire. Knowing where vehicles would be, would not only make maintaining control and ensuring safety easier, but would also protect the range's valuable natural vegetation from free roaming vehicles. A meeting was arranged on the range in question for me to put my case to the conservationists. On arrival I was greeted with banners and a vociferous crowd of about thirty people, who were there to make a noise rather than to listen to anything I might say. German TV and press were evident, but there was no sign of a representative of the *regierung*, who I had expected to be present to support me. Inevitably, the meeting did not go well and a second one was held to no better effect. It seemed to me that the *regierung* chose to stay quiet and let the British take the flak. That range improvement was never completed in my time at Sennelager, although lesser ones were.

While that was a knock back, it left me with funds to address the barracks. An extensive programme to improve heating, washing and toilet facilities commenced, together with inside and outside re-decoration. It was completed in a matter of months, a remarkable effort as the refurbishment involved two battalion-sized barracks, plus accommodation for students at STC's training establishments. All in all, the project improved living conditions for around 2,000 troops. I consoled myself that I had improved the training centre as much as I possibly could.

Domestically, Sue and I were never better looked after than at Sennelager. I had a PA who had previously been a WRAC clerk. Lorraine ran everything –

me included – with a broad smile and frightening efficiency. My driver was Corporal Grimshaw, a long service list soldier who had already completed twenty-two years' service in the infantry, but was allowed to serve longer in a less demanding role. Newmarket House came with two cleaning ladies, who also doubled as waitresses when we gave formal parties and, thankfully, the large lawns were cut as part of the camp's gardening contract.

Sue and I were not used to so much help and decided we should put the Newmarket House crew to the test. Rather than a full-blown dinner party, we thought we would begin with a Sunday lunch. About an hour before the meal, a guest telephoned to say that his Colonel of the Regiment had turned up unexpectedly and could we accommodate the general and his wife for lunch? The answer could only be 'Yes' and the necessary adjustments were made.

Corporal Grimshaw, a Barnsley lad, was to serve drinks. He had never done anything like that before and was as nervous as a kitten. I told him there was only one thing to remember – an officer with an empty glass is unhappy. For the next two years my alcohol bill suffered as a result of that.

The two lady waitresses were almost equally apprehensive, but pre-lunch drinks went well and the animated conversation over the meal's starter indicated that our guests were enjoying themselves. Plates for the main course were laid and our waitresses brought in serving dishes. The Colonel of the Regiment's wife was on my right. Her husband was the general commanding British forces in Berlin and the Villa Lemm, their magnificent official residence there came with about twenty staff, all of them practised professionals. Wendy, one of the waitresses, was waiting to serve her, but the general's wife was in deep conversation with me.

'Oi,' said Wendy, 'the meat's getting cold. Would you like some?'

As Wendy moved further around the table, the general's wife turned to me, smiled and said, 'Have you had your staff long?' How very British!

Elsewhere on the home front, Nick was in his last year at school and had been made Head of House. Tim was enjoying Haileybury and was advancing in rank in the school's Combined Cadet Force. We had a watery holiday at a sailboarding and sailing centre in central France, before moving to canoe down the magical River Lot, with kingfishers as our frequent outriders. Two white cats, Ben and Whizzy, had been dumped on us when we arrived at Sennelager and now we were given a bantam cockerel (Bertorelli, because his black tail feathers matched those in the hat of the 'Allo, 'Allo character), and two bantam and two full-sized hens. They spent their nights in a coup in the rough area, but otherwise roamed the extensive garden. The yolks of their eggs were deep, deep orange and dinner party hosts would often appreciate eggs rather than bottles of wine as our gifts. We were enjoying a particularly happy, expanded household.

My role as Brigadier Infantry meant that in most months, I would pay a formal visit to one or other of the battalions based in Germany. Those visits provided insight into the lives of vibrant military families and although I had criticisms, I was pleased with the professionalism and camaraderie that I was fortunate to witness. Each British infantry battalion had its own character, but without exception, they were remarkably good. Any other army would have been proud to number them in its ranks. What I found fascinating was how little it took in changes to key personnel, to effect mood and performance. Commanding officers were crucial – they set the tone – but like top football teams, a new striker, or the introduction of a constructive mid-fielder, could make a profound difference. In battalions those changes might be a company commander, or the RSM, but the effect was the same. The Brigadier Infantry was a privileged observer and it was a very rewarding job.

One of those formal visits was to 1st Battalion, Royal Welch Fusiliers in November 1991. They occupied Montgomery Barracks in Berlin, where I had served with 2 R ANGLIAN in 1979 and 1980. Part of my visit was a helicopter flight across the former border with the DDR that ran around part of the barracks. Although we knew the Soviets were close to us, it was a shock to see just how adjacent they were. Montgomery Barracks and its Soviet equivalent were separated by only a kilometre or so and a tank training area extended to within a few hundred metres of the British barracks. The helicopter was permitted to over fly the Soviet accommodation complex – it was all but derelict. The Soviets spent their budgets on hardware. Their soldiers seemed not to be a priority.

A direct return flight would have taken us over the Soviet tank training area, but my pilot made a wide detour, having told me that on most days the Soviets blew up munitions that were not safe enough to be moved back to their homeland. We were well past the training area when there was an enormous explosion behind us. The helicopter bucked as the shockwave engulfed it, but the pilot quickly regained control. It was a violent reminder of the extraordinary times that we were living through.

Several months later I was invited to visit a tank range in the former DDR, a few kilometres south west of Berlin. Corporal Grimshaw drove me there and it was fascinating to travel through the previous communist country and get an indication of what its inhabitants wanted most to improve their lot in life. Both Germanys had bulk rubbish collection days when large, unwanted items could be placed at the roadside, ready for municipal collection. In eastern Germany old televisions predominated at the curbside, as the need for better, wider, uncensored communication was paramount. Next came domestic boilers and beds. The ancient boilers were huge and some of the beds were museum pieces, many with rope webs to support straw/horsehair filled mattresses. The rubbish told its own tale.

On arrival at the range I saw the same sort of derelict accommodation that had been evident during my helicopter flight in Berlin. It was unheated. There were toilets, but the only washing facilities were outside stand pipes, each one to cater for about one hundred men. In stark contrast, the tank sheds had elementary heating systems to help maintain the armoured vehicles during the cold winters – a remarkable philosophy of 'look after the tanks and sod the men'.

I was briefed by the Bundeswehr commandant to be told that the range was far bigger than that at Sennelager and was then taken to see examples of its targetry. All of it was operated by mains electricity and I was shaken to learn that the principal tank range, which extended for some 12 kilometres, had all of its target mechanisms hard wired to raise and lower targets as required. The cabling must have cost a fortune. Ranges in the west had their targets operated by batteries, which required constant changing – inefficient, time consuming and costly in man hours, but overall the system was affordable. That difference and the appalling accommodation, underlined the disparity in priorities and starkly contrasted the thinking that had prevailed either side of the inner German border.

Before leaving, I was taken to an assault course to see it demonstrated by a Bundeswehr captain, who looked to be in his upper forties. He made light work of all the obstacles and he was then introduced to me. It transpired that until reunification, he had been a colonel in the Volksarmee and had commanded an air assault regiment. He told me that its war role had been to take out part of the British sector of West Berlin. I had met my enemy and he seemed to think that the British in Berlin would have been a soft target. As he left and I got into my car to return to Sennelager, Corporal Grimshaw whispered, 'Big mouth and an unmitigated shit, sir!' Not much gets past an experienced NCO.

Nick had left Haileybury at the end of the summer term in 1991. His 'A' level results were good and had secured him a place at Birmingham University to read French and English, but not for another year. He was to have a gap year in Zimbabwe. Haileybury had an association with Ruzawi School, a prep school at Marondera, and sent two 'young gentlemen' there every year as teachers' assistants. Additionally, John Rourke, my Technical Quartermaster in Minden, was now quartermaster of the British Army Training and Advisory Team in Harare and Peter Dixon was the military attaché in neighbouring Namibia. Three good reasons to spend a spring holiday in southern Africa.

Our holiday began with a few days spent with John and Joyce Rourke at their stunning hilltop hiring outside the Zimbabwean capital. We took the opportunity to travel to Marondera to see Nick, who, in a matter of months, had matured and adopted a smart, colonial look. He had a spacious flat, full of ancient furniture, which was looked after by his African servant, Ignatious. The flat was adjacent to a dormitory for about thirty small boys who were in Nick's care when not being taught, or otherwise looked after, by the teaching staff. During the day, Nick and the other 'young gentleman' generally acted as assistant games masters. They had landed on their feet.

We left Nick and Zimbabwe to fly to Namibia for an adventure with Peter and Penny Dixon. After a brief stopover at their hiring in Windhoek, we set off on safari in two 4 × 4s. Tim and the three Dixon children, Charles, another Tim and Claire, were with us. We camped amongst the awe-inspiring, red, sand dunes at Sossusvlei – driving off jackals that had been attracted by the smell of breakfast cooking – before moving to the German colonial town of Swakopmund, then north along the desolate Skeleton Coast, before turning inland to the oasis camp at Palmwag with its desert elephants. It was almost too much to take in.

After returning to Windhoek, we took our leave of the Dixons to fly back to Zimbabwe. Ruzawi School was in holiday recess and Nick was free to join us as we continued our trip. We no longer camped. We stayed in a comfortable hotel on Lake Kariba, with hippos on the lawn outside our rooms. Then on to a luxury hut on Fothergill Island, and the fabulous, not-to-be-surpassed Victoria Falls Hotel. The Falls dwarfed those of Niagara and everywhere the scenery was magnificent. Wild game had been on our doorstep throughout and we were privileged to see some excellent native art and dancing. Possibly, the highlight of the entire trip was a visit to the wonderful ruins of Great Zimbabwe. Their colossal size and sophisticated construction exceeded all expectation. What a holiday!

A major focus for the New Year was STC's Centenary celebrations. I was the training centre's fiftieth commandant. Prior to me, there had been twenty-five German and twenty-four British predecessors. The even split demanded that the anniversary be celebrated jointly by both nations. The day of the anniversary fell on 3 July. A memorial plinth, depicting a map of the ranges, was unveiled and then dedicated by our padre before the Bürgermeister of Paderborn, the Bundeswehr Liaison Officer at STC and myself planted trees. My RSM and his Bundeswehr opposite number planted a time capsule and the final act was to re-name six of the roads within the camp – one was 'GROVES DRIVE', which was quite an honour. A week later the Centenary Summer Ball was held and the celebrations were completed in September when a Beating of Retreat ceremony and reception were held. Beating of Retreat is a ceremony dating to seventeenth-century England and its first use was to recall nearby patrolling units to their castle. It has developed to become a musical extravaganza performed by bands, pipes and drums. Our Germans guests loved it and it helped Anglo-German relations at Sennelager recover from the previous year's hiccough over ranges.

After the break-up of Yugoslavia and the Slovenian and Croatian succession from the former federation in 1991, the Muslims of Bosnia and Herzegovina were soon at loggerheads with their fellow countrymen, orthodox Serbs. A referendum, held on 29 February 1992 and boycotted by the Serbs, resulted in an internationally recognised declaration of independence. In response to the declaration, the Serbs mobilised and ethnic cleansing began, forcing the United Nations to act.

Eventually 2,400 British troops were committed to the United Nations' Bosnian operation, which lasted until December 1995, but the lead element in 1992 was 1st Battalion, The Cheshire Regiment, commanded by Lieutenant Colonel Bob Stewart. In late August I was warned that it was all but certain that the Cheshires would deploy, although a firm decision had not been taken by government. The seriousness of the situation was evident because I had just three days to design and set up a five-day training package at STC that would help prepare the battalion for its primary role of providing armed escorts for UN humanitarian aid convoys. The Cheshire battlegroup consisted of the battalion, an armoured recce squadron, an engineer squadron and a composite logistic unit – about 1,800 troops all told. The experience my staff had gained in training 1st Armoured Division for the Gulf War stood us in good stead and we were able to get the Cheshires off to a flying start. They eventually began their deployment in late October.

The rapid training of 1 Cheshire to undertake a particularly nasty operational tour was the last of the short notice upheavals that STC was to experience in my time as commandant. From late 1992, its training establishments and ranges reverted to more normal schedules, but that still entailed the ongoing commitments to train units to serve in Northern Ireland and to help make ready replacement battlegroups to go to Bosnia.

Nick returned from Zimbabwe and after a brief stop in Sennelager, moved on to university life in Birmingham. French and English quickly proved a step too far. By the first half term break he had dropped English. 'Middlemarch' and lyrical ballads were the last straws for him and from then on French had his undivided attention. My brother, Rick, had found his first employment in Birmingham and with his wife Kate, had brought up his family in the city. He was a senior fellow at the university and became acknowledged as a world authority on housing and environmental health. After my father's death, my mother had moved to live in sheltered accommodation only 200 metres from my brother's home. Birmingham had become the centre for my wider family.

The Pompadours had also been on the move, from Minden to Londonderry in 1991. After only a little more than a year in the Province, they made their last move, a return to Meeanee Barracks in Colchester. The Army's post-Cold War drawdown had caught up with the battalion and in October, after 304 years' service to sovereign and country, the Pompadours merged with the 1st and 2nd Battalions.

The final parade was bittersweet. The battalion paraded in front of a huge and emotional crowd of former soldiers, their families, friends and guests. The parade was immaculate, but everyone knew that the end was nigh. At its conclusion, the last commanding officer, Robin Chisnall (one of my company commanders in Minden), asked for and received permission to march off. Before he gave the order, all former Pompadours were invited to form up behind the battalion and march off with them. We 'old boys' outnumbered the serving soldiers by two or three to one. *Rule Britannia* rang out. We marched off with our Colours flying, drums beating, bayonets fixed and our heads held high, even as tears coursed down many manly cheeks.

That night there was a ball, but I remember little of that. What stays vividly in my memory is the church service the following day. It was held at the Regimental Chapel, in Warley, in Essex. The Colours were to be laid up. The church was packed and as a former commanding officer, I was given a place in the choir stalls, with Sue and Tim. The service began and the Colour Party made its entrance and slow marched up the nave to the altar steps. The Escort presented arms and the chaplain came forward. I could not help but think that this was like watching a death, with the soon to be deceased's breathing becoming shallower and shallower. The senior ensign extended the Queen's Colour to the chaplain. He took it and draped it on the altar and turned to take the Regimental Colour. As he did so, the Pompadours died. The Colours were out of our possession. As a unit and as a military family, we were no more.

I do not want to be maudlin about the loss of the Pompadours. The merger has been a great success. The Pompadours brought fresh impetus to the 1st and 2nd Battalions and they were extremely warmly received in their new homes. Their influence remains extant today and the memory of a unique military family is undiminished.

Just over two years had elapsed since Kuwait had been liberated and the rebuilding of its infrastructure was underway. That included the reconstruction of its armed forces and the UK wanted a slice of the lucrative contracts that would result. In April 1993 I was called to the MOD and briefed to head a small delegation to go to Kuwait, to negotiate for the establishment of a British training and advisory team to assist with rebuilding its army. Kuwaiti acceptance of such a team would bode well for sales of British weaponry and equipment.

My team consisted of myself and two lieutenant colonels – one was an artillery officer, the other a logistician; both had served with Arab forces and spoke reasonably good Arabic. We were met at Kuwait City's airport by the British military attaché and were taken to the Embassy for a locally orientated briefing. We then moved on to our hotel, each with a bottle of spirits in our baggage (courtesy of the military attaché) to help us through the 'dry' days ahead. I was given a suite of rooms near the top of a skyscraper building. The suite was vast. It had an opulent sitting room, an area for a conference table and chairs, bedroom, dressing room and a gold plated en suite (horrendous). The bed was so large that lying across it, with my arms fully extended above my head, I was nowhere near touching its sides. I told the military attaché that the suite was completely over the top. He quickly informed me that we were being closely observed by the Kuwaitis and that status was hugely important. The Americans and French were in town, after much the same things as we were, and, if I was to have credibility, my accommodation should at least match theirs. Dirty work, but someone had to do it.

The following morning, the team was met in the hotel's foyer by our Kuwaiti liaison officer. He was a major, a fast jet pilot, who had trained in America and he spoke excellent English. In passing, he also mentioned that he was a prince. There are lots of princes in Kuwait, but it did no harm at all to have a liaison officer from a leading family. We set off to begin a demanding schedule of meetings with influential senior officers, all of them in charge of key components of the Kuwaiti military. We would have three or four meetings each day for three days and then a day to write a report that had to be ready on our return to London.

At the first meeting I was kept waiting, in an outer office, for about ten minutes. Another five minutes ticked by and I told the liaison officer that it was unacceptable. He remonstrated with an aide and then walked through a door into the senior officer's office. Raised voices were heard briefly and then the senior officer appeared to usher my team into his office. It seemed that our prince had standing and I had not lost face by merely accepting a wait.

Once in the office, tiny cups of tea were served by a waiter. During the previous evening, my two lieutenant colonels had helped me brush up on my few words of Arabic. The fact that I was able to greet the senior officer in Arabic and accept my tea with a shukran (thank you) did not go amiss. The competent Arabic spoken by my wingmen ensured that we made an excellent initial impression and the meeting was highly successful. The senior officer wanted British assistance, more of it than we could have reasonably expected.

That provided something of a blueprint for the meetings that followed. With their past experience of Arab forces, the two lieutenant colonels understood Arab sensitivities and that manners counted for everything. I was advised of various rituals (like shaking my cup between thumb and forefinger to indicate that I had had enough tea – simple enough, but it brought smiles of approval) and not to dive into negotiation, but to allow some considerable time for polite conversation, if possible, letting the Kuwaiti officers broach the subject of British support. We were careful to ensure that our liaison officer had time for prayer and, on one occasion, joined him, kneeling on the floor at the back of a prayer room in a military headquarters. We were made very welcome. At the end of the three days, we had secured requests for a training and advisory team that was, at least, three times the size of the one that had been envisaged. We tried not to be too pleased with ourselves and settled down to write our report.

In the middle of our drafting, the liaison officer appeared at our hotel. We were surprised to see him because his work with us was over. He said he would be honoured if we would join him for a meal that evening and we were pleased to accept. It was a traditional Arab meal, in a very smart restaurant and, again, I was glad of the prior tutoring of my experienced off-siders. When the meal was over, there was another surprise in store for us – we were invited to meet the liaison officer's family at his parents' home.

The house was large and plain from the outside, but inside it was generously proportioned and ornately decorated. We were met by patriarchs – the liaison officer's father and four of his uncles. They were amongst the most dignified men I had ever met. We were shown into a room with colonnades and a mezzanine floor. Behind a fretwork screen on the mezzanine, were the adult women of the family. We settled ourselves on cushions and knew that we were being closely observed by those in front and above us. The conversation that I had now learned to expect, went on and on and deepened as the gentlemen probed our military, but particularly our family lives. We responded with questions about our hosts. At well past midnight I thought that we should leave, but playing cards were produced. We picked up the game quickly and more hours flew by. We left in the early hours of the morning having had a most unexpected, enlightening and enriching encounter. 'Alhamdulillah' (Thanks be to God).

The following morning the team was taken to the British embassy for a courtesy call on the ambassador before leaving for the UK. He congratulated us on our achievement and was particularly impressed that we had been invited to an Arab home, an honour that could take years to secure, if an invitation was ever extended at all. In London, our report was warmly received and within months the training and advisory team was in place.

On return to Sennelager, I was met by a worried Sue. Barely eighteen months after the Berlin Wall had been breached, and as a consequence of the *Options for Change* initiative, the British Army had already begun the process of drawing down. Two of the local cavalry regiments were to merge and their royal Colonels-in-Chief were to be present at ceremonies to mark the merger. Prince Charles, Colonel-in-Chief of the 5th Royal Inniskilling Dragoon Guards, was to stay with the local brigade commander. Sue and I had been asked to accommodate the Duchess of Kent, Colonel-in-Chief of the 4th/7th Royal Dragoon Guards. I could not see a problem, but Sue told me that a recce party had deemed it necessary for us to give up our bedroom for the Duchess. I could barely believe what I heard. The quirky layout of our home meant that our bedroom was about four metres closer to the bathrooms than the guest bedroom. Exactly the same muddled, sycophantic thinking that had marred Princess Diana's visit, was being repeated. I lost no time in making clear that while it would be an honour to have the Duchess to stay, there would be no changing of bedrooms.

That was the last we heard of that and the Duchess duly stayed for two nights. After she had taken tea with us on her arrival, we saw very little of her, other than when she came to and left the house – we were required to be present at those times. She took breakfast in her room and all the other meals were official ones, to do with the merger and were held elsewhere.

Her entourage were a pleasure to entertain. It comprised four people and that stretched our accommodation to the limit. The lady-in-waiting took Nick's bedroom – she had a teenage son and was quite at home. The close protection officer – an inspector in the Metropolitan Police – slept in the office that fortunately had a toilet and wash basin immediately adjacent to it. The maid slept in a small bedroom on the first floor, close to the Duchess's room and the hairdresser was banished to Tim's room in the attic.

We received very little briefing for the visit and on the first morning had laid breakfast for ourselves, the lady-in-waiting and the inspector in the dining room. Not knowing how the hierarchy of a royal household works, we had laid places for the hairdresser and maid in the kitchen – we had probably watched too much *Upstairs, Downstairs* in the past. The lady-in-waiting immediately re-laid them with the rest of us, saying that most hosts got that wrong and that the four were a team – there was no differentiation. We were very happy with that, especially when we learned that the hairdresser owned a chain of West End establishments and was not just a dab hand with a pair of scissors, spray and a comb. He was also a keen fly fisherman and when I told him that I was too and had my own trout lake, he was impressed. Soon after daybreak on the second morning, he disappeared with my rod, to fish successfully in the Commandant's lake. His smile meant that I believed him when he said, 'I wish all visits were like this.'

Sue and I attended the merger parade and a ball and the visit was quickly over. On the morning of her leaving, the Duchess was gracious in her thanks and we were presented with a signed photograph. About five hours later, I turned on the television to watch the FA Cup Final, played at Wembley. There, in the Royal Box, was the Duchess ready to meet the teams and present the trophy. I have met many of the royals over the years and have my criticisms of some of the people who surround them, but I often have had cause to think that their duties are more demanding than their detractors would have us believe.

I had had little contact with my Regiment since the demise of the Pompadours. Out of the blue I got a telephone call from General Sir John Akehurst, the Colonel of the Regiment. He told me that I was to become a Deputy Colonel of the Regiment. I was delighted with the unexpected honour and looked forward to the visits to the battalions and old comrades' associations that the appointment would bring. Because the 2nd Battalion was based in Celle, less than two hours' drive from Sennelager, I was able to see them regularly, but I had also been given a particular responsibility for the 7th Battalion and they had companies scattered across the east Midlands. They and the old comrades would have to wait until I returned to England.

I had now been in post at Sennelager for approaching two and a half years and my next posting was due. I was tempted by command of a Territorial Army brigade and the quarter that went with it – Dover Castle! However, General Sir Charles Guthrie had taken over as Commander-in-Chief BAOR and he intervened. I knew him well, first from my time at MOD, then at BBGT when he commanded 1st British Corps and, latterly, as Brigadier Infantry

(BAOR). 'Go and sort out recruit training,' was his advice and I elected to become Commander Initial Training Group. That meant a move back to the UK after some ten years in Germany, all of which had been kind to me.

I was keen to know who would take over from me as Commandant at STC – the post of Brigadier Infantry was discontinued on my leaving and the Army reduced in size. I should have guessed who my replacement would be – Alan Behagg! He was running the British Army Training Unit Suffield, in Canada and could not be released until the training season ended there in the autumn. That required me to have two jobs for a couple of months, one in the UK and the other in Germany, but that was not a problem. *Auf Wiedersehen Deutschland.*

CHAPTER – EIGHTEEN
SWANSONG AT UPAVON

THE TEN YEARS in Germany and four different quarters that had allowed for some expansion of our possessions, meant that we had a lot to move to the UK. The cats went first, collected by people from a cattery, who made a good living from transporting service families' pets from Germany and then looking after them in the UK for the six months of the quarantine period. The chickens went next, locally, to another brigadier and his wife and then we got down to packing. During our time in Germany, the Army had changed its rules – the MFO boxes had gone and in came removal companies making moving so much easier.

After an overnight crossing from Zeebrugge to Felixstowe, we drove straight to Ipswich. Our cars were registered with British Forces Germany and in response to the IRA threat to service families, we now had what looked like German number plates. We needed to re-register the cars at the local office of the then Driver Vehicle Licensing Centre and to purchase British plates. All went well until an official spotted that the externally fitted, reversing light on the Daihatsu was on the wrong side for driving in the UK. 'You'll have to get that changed before the car is legal' was his uncompromising advice. We tried a local garage and got nowhere. Our next stop was scheduled later in the morning, at Haileybury, where Tim finished school on that same day and needed his stuff collecting. As we were not going to drive in the dark, we risked driving the illegal car.

Tim was not 'flavour of the month'. Days before we had left Sennelager, he announced that he was not going to university immediately, but would take a year out instead. He had no plans for what he might do and that needed some attention.

At Haileybury we crammed his stuff into the cars, only to be told there was more of it at a friend's house a few miles away and his popularity rating took a further dive. After more rearranging, pushing and shoving, the last of his possessions were somehow loaded aboard and we were free to drive the last leg of the journey to Upavon in Wiltshire. Tim departed on holiday, inter-railing around Europe – alright for some!

The River Avon runs south from Devizes to Salisbury and then on to the sea. At Upavon the river is joined by the A345 and, side by side, the river and road bisect the Salisbury Plain Training Area. The village of Upavon is divided by the crossroads of the A345 and the A342, which climbs steeply out of the Avon valley to continue eastwards towards Andover. On top of the first, windblown, barren hill lay RAF Upavon, our new home.

The gate guard waved me down as I approached the camp's barrier and I was asked to identify myself and 'sign in'. My name was recognised and I was given a large, brown envelope. It contained a welcoming letter and the keys to the residence we had been allocated. Sue and I drove through the camp to a quiet road of pleasant quarters. Trenchard House was significantly bigger than the others. It was probably a 1930's build with three storeys; the upper one was the roof space, but dormer windows indicated that it had been designed to accommodate bedrooms. A narrow, single-storey block on one side of the house had a series of small rooms and linked to a garage. The garden was sparsely planted, private and generously proportioned. It all looked promising.

Having struggled for several minutes to open the front door – the lock had been installed upside down – we crossed the threshold of our new home. We looked into a generously large kitchen, but the work surfaces were of chipped, scuffed, red Formica; some of the doors of its worn cabinets were misaligned; at the far end of the room an ancient, coke-fired Aga, with badly pitted hot plates and burn marks on the lino that surrounded it, seemed to stare malevolently back at us; two mismatched, dated cookers stood next to it; and above them, a series of bells hung silent, waiting to be rung to call for servants who did not exist. It was awful. My strong, tired, resilient wife cried.

Tea came to the rescue and we unpacked most of what was in the cars. Then we turned our attention to the 'get you in' pack, made down our bed, sorted out its pots and pans and the groceries that we had bought on the way. We were set for the night, but had no time to feel sorry for ourselves as we had to travel to the Midlands the following morning, to pick up Sullivan – a new cat.

When we had moved to Germany in 1982, we decided that it would be wrong to take Marks and Spencer with us. They had gone to live with Nicola, our mutual schoolfriend. She had loved them and kept them until they died, but she was about to move to live in Majorca and wanted her favour returned. It was a request we could not refuse, so little, black and white Sullivan, with his stiff gait and extremely sharp claws, came to live with us, joining Ben and Whizzy after their quarantine.

Sue and I were taking up residence on a RAF station because, like the other services, the RAF had to slim down and only shortly after our arrival, the station would be handed over to the Army. It was a big wrench for the RAF because Upavon was the first base built for the Royal Flying Corps in 1912. Notwithstanding that, the RAF had been realistic about spending money on a place it was to give up. Maintenance and updating had been low priorities. The Army had to spend a huge amount to bring the station up to scratch, including the demolition and rebuilding of some junior ranks' married quarters and the replacement of a manual, plug-in telephone exchange that in all probability, dated from pre-World War II.

The Army built a new headquarters for the Adjutant General at Upavon. Other headquarters would also move to occupy the station's existing buildings. One of those headquarters was the Inspectorate of Doctrine and Training and I would be subordinate to it in my new role. However, before I attempted to get to grips with recruit training, I had a first duty to perform as Deputy Colonel of the Regiment and that entailed a trip to America.

The re-enactment regiment, the 10th of Foot of America, had their Colours made by the Royal School of Needlework and I was to represent the Colonel of the Regiment at their presentation. I turned the opportunity into a private holiday and Sue and I flew for a second vacation on the American east coast, to stay with family and friends after attending the presentation. At the parade the new Colours were magnificent. Red coats, white breeches, powdered wigs, swords and muskets made it a throwback in time. In the evening, a celebratory ball, held in a period building in Salem (of witches' fame), was just as grand. The 10th knew how to throw a party.

Refreshed and back in the UK, I was happy to get down to work. For my first few days as Commander Initial Training Group (CITG) my headquarters was co-located with HQ UKLF at Wilton, some 20 miles south of Upavon. As the RAF moved out of Upavon, we moved into our new offices. I had a colonel as my second-in-command; two majors and two captains as staff officers and a lance corporal as my driver. The remaining staff, comprising another wonderful PA, Jackie, two budgetary officers and three clerks, were all civilians. It was a tight knit little group.

The Initial Training Group was scattered across Great Britain. The Group comprised five Army Training Regiments (ATRs) located near Edinburgh, Lichfield, Bassingbourn in Cambridgeshire, Pirbright in Surrey and Winchester. Every ATR was commanded by a lieutenant colonel, but the regiments varied in size and each trained recruits for different corps and regiments e.g. the Armoured Corps, Infantry, the Royal Artillery, the Royal Logistic Corps and the Adjutant General's Corps. In the 1990s, only the Infantry and Armoured Corps recruited solely male soldiers. All other regiments and corps included both sexes in their training companies, both as recruits and as instructors.

Within the first couple of weeks I had visited the five ATRs. Each corps and regiment sent extremely good officers and NCOs to instruct their recruits – it was in their best interests to have their soldiers as well trained as possible. With regard to the instructors, I was impressed with what I saw. The recruits were a different matter. I had been allocated this job to reduce the high drop-out/ failure rate in training, but that was to prove difficult.

Of all the young people who applied to join the Army, only about fifty per cent made the grade to start training. Some failed on medical grounds, others were insufficiently fit, but a significant proportion lacked commitment. Having expressed an interest in joining, many could not be bothered to turn up for assessment, or try hard enough to pass the preliminary tests. The Army would not have wanted those people anyway, but it left a reduced pool with which to work.

On arrival at an ATR, recruits had to adjust quickly to an environment hugely different from the ones they had known. Just a few had to have shoelace tying and knife and fork lessons – Velcro fasteners and pot noodles, eaten with a spoon, have a lot to answer for in some youngsters, often those from disadvantaged households. And 'wash behind your ears' (or simply to wash) was now not just a parent's forlorn hope. It was an order. Ignoring it was not an option.

Once training began, recruits' physical fitness, stamina and resolve were sorely tested. The vast majority of them were fit and strong enough to undergo training successfully, but around twenty-five per cent of those who started training found the going tough. For instance, virtually all could run three miles in the time required, but if asked to repeat it (or something equally demanding) on following days, they often found that stamina let them down and they had yet to develop the resolve necessary to overcome testing circumstances. Boots caused constant problems in the first few weeks of training. Having worn trainers almost exclusively up until that point, the feet of many recruits were too soft to wear boots throughout a working day. 'Pelvic stress' was a particular problem for many of the smaller female recruits. A marching pace is of 30 inches and comfortable enough for men, but it is literally a stretch for small women. A lot of a recruit's working day involved marching from one place of instruction to another and, in addition, there were drill periods to negotiate. During all that activity, small women would be over striding and that resulted in pulled muscle in their pelvic floors – a really painful condition that needed rest to cure and removed them from training until they were well again. Unsurprisingly, the physiotherapy department was amongst the busiest of places in every ATR.

Another problem worthy of mention is that of the male role model. Just as Sergeant Mason had spent extraordinarily long hours helping me and the other officer cadets of 5 Platoon at Mons and had consequently gained our respect and admiration, the same was true for the training corporals at the ATRs. They had eight to ten recruits to husband through the recruit syllabus and spent long hours doing it. Often, they were the first male role model that some recruits had encountered – many came from broken homes. That was a positive for male recruits, who wanted to emulate their instructors. For a few female recruits it could become a complication, particularly if they developed crushes on their most readily available instructor. Fit, testosterone-charged, male JNCOs needed to know how to deal with that situation and that was where our female instructors were invaluable. They were strong advocates of treating male and female recruits equally (everyone endorsed that), but they did advise on matters like how to react when a young woman might 'try it on' with a few tears, or when the debilitating effects of a difficult menstruation meant that a more sympathetic approach should be adopted.

When I took over as CITG, the training wastage rate wavered around twenty-five per cent. Any small variations tended to reflect the quality of the cohorts under training. The first five weeks of training were crucial and accounted for a disproportionate amount of the recruits who fell by the wayside. During that time recruits could simply opt to leave the Army without penalty. I gave my second-in-command, Colonel Joe Gunnell, responsibility for scrutiny of the training syllabus, in particular to see if it could be tweaked to lessen the steepness of the initial learning curve. All senior leaders were aware of the need to communicate best practice and my bi-annual meetings of commanding officers were vital forums for the free exchange of ideas. We tried our level best to secure a significant and lasting improvement and I would like to report that we were successful in our endeavours, but try as we might, we never got the wastage rate much below nineteen per cent. Most of the time it remained stubbornly in the low twenties. I have concluded that in the 1990s, and possibly more so now, the transition from civilian to Army life is so great that a significant number of young people will struggle with the physical and mental demands it brings. Any diminution of training standards would leave young soldiers ill prepared for the role that lies in front of them and lower skill levels would increase the likelihood of casualties – a difficult dilemma to solve.

With the Army's arrival, RAF Upavon was immediately re-named Trenchard Lines. It became a building site as construction began on the Adjutant General's headquarters and much of the rest of the estate started to receive long overdue maintenance.

In the first week at Trenchard House, I trimmed two large, ornamental bushes that had grown sufficiently large to obstruct access from a rear door to the garden. The reduction of the first bush went without incident, but as soon as I attempted to cut back the second, I was surrounded by angry wasps. When the rodent control officer removed the nest, it was the size of a beach ball. The sitting room fireplace smoked because crows had built a large nest in the chimney. The upper storey had two spacious bedrooms, one at either end of the building and a third small bedroom in the middle. None of them had heating. A cupboard in the small bedroom opened onto an attic space that contained ancient pipework and an equally ancient cold water tank, all were lagged with straw. We asked for the Aga to be removed and for the rest of the kitchen appliances and cupboards to be inspected. During the inspection, asbestos was found in the boiler room, which resulted in Trenchard House being included in the camp's upgrade. By the end of eighteen months intermittent work the house was fully restored and Sue was confident that she knew most of the workmen in Wiltshire.

With his inter-railing over, Tim came to live, or at least be based, with us.

Both Sue and I agreed that he had to find work and there followed a series of widely varying, unsatisfactory employments in different parts of the south of England. Eventually, he struck lucky.

The loss of Crown immunity meant that the small sewage works at Trenchard Lines had to have a lifeguard present while the man who operated the plant went about his work. After his security vetting was completed, throughout the summer lifeguard Tim was paid to sit under a tree and read a considerable number of books. He had another break when the opportunity came to work with the camp's maintenance team. He joined a long-established team of three elderly men, so set in their ways that their day-to-day interactions would have made an excellent TV or radio series of the *Steptoe and Son* genre. Most evenings Tim (an excellent raconteur) would regale Sue and me with everyday stories of the maintenance team that easily eclipsed those of *The Archers*. Tim also had an area of sole responsibility, a hangar full of the war reserve of toilet rolls. That, when combined with his previous job in the sewage works, made a novel talking point on his CV! He remained a maintenance man until he left to go to university, at Goldsmith's in south London.

Sue had qualified as an independent financial adviser and had remained in touch with Jonathan Hall-Tipping ever since our return from America. Jonathan had moved from stockbroking to become an equally successful manager of private portfolios. He wanted to do the same thing for Americans living in the UK, who had retained portfolios in the States. As a result of that aspiration, a new UK-based company, Westbourne, was born and held its inaugural seminar for prospective clients in February 1994. It was a full-time job for Sue as she worked to get the fledgling company up and running.

The end of the MOD's financial year is 31 March. A windfall of money always became available in the weeks before that date, because high level budget holders husbanded their cash against unforeseen circumstances and with the year end in sight, they needed to spend it if their budgets were not to be reduced for the following year. Anyone who had a short term project that could be concluded by 31 March, was almost certain to get the cash they required. My logistical staff had a list of maintenance tasks ready, but welcome as their funding would be, it was not significant. My budget manager, Mary, had an excellent idea that would secure some immediate funding and put us in a much better position to improve facilities in the next financial year. She asked what the most urgent shortfalls in the ATR's facilities were. The top priority was for a much bigger gymnasium for ATR Lichfield. Mary applied for a modest amount that would enable an architect to draw up plans for a modular building. Those plans could easily be produced before 31 March. She added provision for a concrete base for the building to our main budget application for 1994/95. With plans and a base in place, when the 1995 windfall arrived, a modular build could be completed in only two to three weeks. As a result of Mary's good work, the 'Groves' gymnasium opened at ATR Lichfield in March the following year. The name was mine; the credit was Mary's. We repeated the process the following year and ATR Pirbright also got a 'Groves' gymnasium.

1994 was a relatively quiet year for me. The recruit training regime was settling down well. I had a good staff and the commanding officers of each ATR did not need me peering over their shoulders as they produced good soldiers and wrestled to reduce training wastage. I was far from being at a loose end but I did have capacity to spare.

A fellow officer from Ballykinler days, Trevor Veitch, was extremely keen to preserve Royal Anglian history. Even though only thirty years had passed since the Regiment's formation, vital information, artefacts and ephemera would be lost if their collection was not actively pursued. That meant establishing a Royal Anglian museum to house and display whatever we managed to assemble. Trevor has never lacked nerve and his answer to that problem was to turn up, unannounced, at the Imperial War Museum (IWM), Duxford and ask for some museum space. He explained that Duxford marked the epicentre of the nine counties from which the Regiment recruited and that we were natural partners. The element of surprise worked and he was given space in the Land Warfare Hall, rent free, on condition that our museum matched the standards of the IWM. Trevor now needed help from his mates and I was amongst those roped in. For the next thirteen years I was either a trustee, or chairman, of the museum's governing board. It was fascinating to work with the IWM's curatorial staff, none of whom had military experience, but who were generous with their time and could point us in the right direction with things like preservation, accession and display. We made a good partnership and an excellent museum has resulted.

My interest in football had also not gone unnoticed. I was invited to join the board of the Army Football Association (AFA) and I was pleased to accept.

I found little to do at the AFA as football was widely recognised as the best organised of all Army sports, but the Infantry's representative team was in the doldrums, so I took it on. Football is the soldiers' game, but because most officers prefer rugby, many did not give it the attention it deserved. I wrote to all infantry commanding officers in the UK asking them to take a personal interest in the representative team and to send their best players to a trial in Tidworth. By the end of the afternoon of the trial, the Infantry had talent that could carry all before it. Such was the quality of the team that in the next two seasons, two players were bought out of the Army to play for professional clubs.

My preoccupation with Infantry football was to be my undoing – or so I thought initially. I missed an AFA board meeting and when I got the minutes, I found that my colleagues on the board had unanimously voted me to be the first president of Army Women's Football. Thanks mates!

It turned out to be a great experience. I appreciated that there was a job to be done and accepted the post with the proviso that the women got equal status with the men. I was delighted with the support I received. The trainer of the Army men's team became the trainer of the Women's XI. He had female assistants, who he mentored and a first class female physio from one of the ATRs. A male secretary, with wide experience of Army football, was appointed to organise and administrate the team and the women's home ground was the Army Stadium at Aldershot. From the outset we were a going concern.

A trial was organised at ATR Bassingbourn that had a lot of pitches. The trial had been advertised throughout the Army in UK and we were confident that a couple of hundred players would turn up. In the event, we had double, if not treble that number. The Army Physical Training Corps' instructors produced order out of chaos and by the end of the day we had a squad of potentials for the Army team. We also had enough evidence to justify organising women's football at lower levels. It was the best of all possible starts.

The women's representative team had their first match early in the 1994-95 season against Red Star, then the women's team of Southampton Football Club. I thought I ought to have a word with the girls before kick-off and addressed them, saying what I expected of them now they had the honour of representing the Army and that although they were unlikely to be as skilled as their opponents, their stamina and fitness should count. I asked for any questions and there was silence, eventually broken by the captain. She said 'Are we taking prisoners, sir?' I had no doubts about the combative qualities of the soldiers in front of me.

The match resulted in a well-deserved draw and the team quickly went from strength to strength, even attracting attention from the England coach. Now women's football in the Army is a mainstream sport. Its president, secretary and coach are all knowledgeable and able women and I am proud to have played a small part in its initial success. I am still connected to Army football as a life-time vice president.

We had two family milestones to celebrate in 1994. Nick was twenty-one in the February – where had that time gone? He secured an Erasmus grant to study for a year for Besançon University in France and I drove him there for the beginning of the autumn term. His lodgings looked fairly dilapidated to me, but he was happy and remained that way throughout his successful, enjoyable year in the Jura.

Another case of *tempus fugit* was our 25th wedding anniversary that arrived in September. Prior to that we had holidayed in Rome, staying with Paul and Alison McMillen. Paul had been selected to study at the NATO Staff College there and he had generously extended an invitation for us to visit. It was a round of visiting one amazing venue after another: the Colosseum, the Pantheon, Circus Maximus, the Vatican and St Peter's, the Appian Way, the Catacombs and so many more. In between sightseeing, we ate, drank and laughed a lot.

The Services continued to draw down. Redundancies were in the offing and I knew that even if I avoided them, I would only have one posting left before I reached retirement age. A last posting would be unlikely to be rewarded with a good, challenging job and I was not keen to become a bean counter. I began to consider redundancy seriously and thought an Open University MBA course would help me identify my transferable skills. I applied for a military grant to cover the costs of the course and received funding for the first year. That seemed to be a good initial step towards leaving the Service. Then I took the plunge and applied for redundancy. I would not know if my application had been successful until the following April.

Christmas was spent with the family in Upavon and when the holiday period was over, I returned to work. On the first morning, my driver asked if we were to meet for our normal midday run around the airfield's perimeter. I agreed and we duly met at the camp's gate. When not in public, I used my driver's first name. In the seventeen months she had been with me, I had probably spent as much time in her company as I had with Sue. Lisa was now part of the family. She was a half-marathon runner and very competitive. I could no longer keep up with my young, male staff officers, so running with Lisa was pleasant and it had an edge. Lisa's mantra was 'One day I'll beat you'. 'Not this side of the grave,' I told her. The run went badly for me. Lisa was hundreds of metres ahead of me at the finish. She asked what was wrong and I had no idea, apart from the fact that I could not get my breath and felt extremely weak. 'That doesn't count,' said Lisa. 'Same again tomorrow, sir. No messing. We've got to sort you out.' I agreed and recovered quickly. The following day's run produced the same result and I felt dreadful, but after a shower and a meal I was back to normal.

The next morning, still in my dressing gown, I went downstairs to make tea and to feed the cats. I remember bending over to put the cats' bowls down and woke up a little later with the unconcerned cats happily eating their food around me. I had fainted. After sitting down for a few moments, I carefully made my way upstairs to tell Sue and to get back into bed. Sue took over and called the camp's doctor who came to the house later that morning. He examined me and took blood samples. The results showed that I had lost almost all of my white blood cells and I was immediately referred to the RAF Hospital at Wroughton, only a few miles away.

There I underwent a series of tests without any conclusive result. Eventually, I was given a barium enema. While I was still strapped to an examination table, inverted, and with my intestines full of the barium mixture, I noticed the radiologist hurriedly leave her work station. Only a minute or so later she came back, accompanied by a doctor. They conferred rather earnestly, looking at a screen. 'That doesn't bode well,' I told myself. An orderly released me from the table and showed me to a loo to sort myself out. I was then taken to a small room to await the doctor's diagnosis, but was so uncomfortable that I could not sit down, or remain still. When the poor man came to give me the news, I was still walking around the room, trying to control my tummy.

'You have bowel cancer, sir, and it looks as if it might be malignant.'

'OK,' I said. 'OK, but just let me get back to the loo.'

'Have you taken in what I have just told you?'

'Yes, yes, but if I can't get to a loo quickly, I'm going to fart my way to the moon!'

That might not be the normal response by someone being told that they have cancer, but some things take precedent.

Telling Sue was not easy, nor was telling Tim. Eventually, I was able to speak on the phone to Nick in Besançon to give him the news. He was his normal phlegmatic self. 'You'll be alright, Dad,' he reassured me in his usual serious tone and said very little more. His absolute faith was a comfort.

I was operated on, at Wroughton, five days later. The surgeon, a group captain and professor of surgery, informed me that the cancer had been caught early, that it was malignant, but he thought he had removed it all. He told me that I had a fifty/fifty chance of surviving for another five years. I resolved to do all in my power to do better than that. Within a week I was able to leave hospital and walk 300 metres, along a maze of corridors, to my car and that was testimony to the efficacy of modern surgery and the excellent nursing that I had received.

I started a course of chemotherapy almost straight away. It was carried out at the British Military Hospital Greenwich, in south east London, two hours' drive away from Upavon. There were advantages to going there. It had a first-class oncology ward and the hospital attracted top, London based, civilian specialists, including an oncologist. The attraction for the civilian specialists was that if the military patients agreed, new, experimental treatments could be tried on them. The patients were ideal – young, fit (apart from the bit wrong with them) and combative; eager to make the effort to get well. Added to that the hospital was quiet compared to its NHS equivalents and the competent military doctors and nursing staff had time to devote to their patients.

My specialist told me that he was probably one of the three best oncologists in the country. He added, 'But you are the world's expert on Colin Groves. You must tell me exactly what and how you feel and when you feel it. "I'm OK doc" will not do. If you do all that then we will form a partnership to beat your cancer.'

We shook hands on the deal. He then prescribed my chemotherapy regime – a combination of toxic drugs, to be administered over a two-hour period for five consecutive days, one week in every four. I agreed to try a new drug.

My first chemo session was a frightening affair. I was given a bed in a small, private room and the drip was set up. After a strong, major vein had been identified, a cannula was inserted above my left wrist and the drip was turned on. I was left on my own. Only a few seconds elapsed before I felt a bubbling sensation in the carrier vein. I could trace exactly where the chemotherapy drug had reached and the bubbling was so pronounced that it put me in mind of an old advertisement for a toilet cleaner 'Harpic cleans around the bend'. What was going to happen when this stuff reached my heart! Alarmed I pressed the bell and an orderly explained that the sensation would diminish. I asked to see the ward sister and told her, in no uncertain terms, that patients should be warned of the likelihood of a disquieting reaction in their bodies and be told not to worry about it. She said that I was unusual in that my reaction to the drugs had been extreme, but she certainly would look into it. It might be the new, experimental drug.

After the transfusion, Lisa drove me home. I had elected to travel to the hospital on every one of the five days so that I could spend the nights at home and do a little work in my office in the mornings. The first transfusion made me feel as if I had a cold coming on. By day two I felt the onset of 'flu; by day three I had an aching body; day four and I was really groggy; and after the fifth transfusion I would travel home, slumped in my seat and would spend most of the weekend in bed. The following week a reverse process took over. Day by day I felt better and was right as rain by Friday. The MBA course proved to be a boon. Studying kept my mind off cancer and masses of reading and the bare bones of several essays were formed in my small room in Greenwich, while the drip dripped away.

The second round of transfusions brought a surprise. I was well into my recovery week when I looked into my shaving mirror to see that my face had turned purple. I immediately rang my consultant. He asked 'How purple is this purple?' I told him it nearly matched my Pompadour lanyard. 'Get yourself to London immediately.' I did as I was asked and we met at the Charing Cross Hospital. Two representatives from the drug company were with him and I was taken off the drug at once and another was prescribed.

I was due to have chemotherapy for a year, but after six months my body rebelled. I was sick and had diarrhoea. That was usual after every chemo session, but this time it went on and on. I lost a lot of body fluid and part of my stomach wall. I was rushed to RAF Wroughton to be rehydrated. Apparently, I looked so ill that Sue thought I would die. However, just as

Nick had foreseen, I recovered. It took time to rebuild my strength and chemotherapy was discontinued. Thereafter I was monitored every six months for five years and my civilian specialist remained available to me during that time. A review regime is still in place, but a minor discomfort every five years is a small price to pay for remaining above ground.

I am one of the lucky ones. Twenty-five years later and I am in rude health. Early in 2020 I drove a husky team across the Arctic tundra in a temperature of minus 28 degrees centigrade. It was one of the highlights of my life and a testament to the wonderful treatment I received from the RAF, my specialist and the Army. It gives me considerable pleasure to record my sincere thanks here. I should also thank Lisa. The early discovery of my cancer was largely due to her and our runs together.

While I underwent chemotherapy, I took advantage of the Army's resettlement programme for senior officers. On a course run by the bankers, Coutts, CV writing was a given, but it also offered excellent advice on how to transition to civilian life and how to decide on which field(s) of endeavour might best suit. I began my job search. It was a frustratingly slow process.

By the autumn I was fit enough to travel abroad for a holiday. Both Sue and I needed one after the rigours of the first half of the year. We flew to Istanbul, to stay in the Pera Palas hotel. It had been a favourite of Agatha Christie, set on a hill overlooking the Bosphorus Bridge and the main railway station, where the Orient Express would have delivered Agatha and its cargo of elegant ladies and gentlemen at the end of their journey across Europe. Little had changed in the hotel since those times. On stepping inside, one entered a forest of polished mahogany. The lift – with no interior doors – rose with a gentle elegance to the upper floors, controlled by a bell-boy in a pillbox hat. Our bedroom had a faded sophistication and the bathroom's pipework looked as if it had been taken from the control room of a swimming pool. The eccentricity of the place added to the ancient charm of Istanbul.

We simply did the tourist thing and immersed ourselves in the Blue Mosque, St Sophia, Topkapi, Roman cisterns and markets. We also took a ferry ride beyond Cylla and Caribdis, to be caught in a violent storm on the return journey – it was the only time that I had ever to don a lifejacket on a ferry. Other than that, it was a much gentler holiday than those we normally took, just right for us at the time.

Earlier, I focused on the problems that my organisation had to contend within training the less robust recruits. It would be remiss not to give balance to that picture with the effort made and success achieved by the remainder. It is my unshakeable conviction that there is absolutely nothing wrong with the bulk of young people today. Many are gifted and the vast majority are capable of achieving much more than they suspect on their entering adult life. For those who join the Army, undergoing recruit training can be like pressing an accelerator connected to their potential. After five months training many are barely recognisable to family, friends and their former teachers. They stand taller; most are physically bigger; they are confident in themselves and of the others around them; they are proud and their broad smiles tell of how far and how well they had travelled in a short time.

At the end of recruit training, passing out parades are impressive events. The recruits dress in their best uniforms and have worked tirelessly to perfect their drill. A band is always present and a senior officer inspects the parade and takes the salute. I was privileged to do that quite often.

One passing out parade stands out in my memory. It was at Bassingbourn, shortly after my chemotherapy had been curtailed. I was inspecting the parade when I came to a stocky, young Royal Engineer with tears rolling down his cheeks. I asked him what was wrong.

'This is the first important thing I've ever managed to do,' he said.

I told the officers accompanying me to close together to screen the young soldier. 'Get rid of the tears. Give me your rifle and take this.' I gave him my handkerchief.

We let him have a moment to compose himself. 'Keep it,' I said when he offered to return it. 'Now shake my hand and take your rifle. Many, many congratulations trained soldier,' I said and meant it as I moved on.

Reflecting on that incident, I am still pleased to have met that young Sapper and I am proud of him, his fellow recruits and of the instructors who wrought such a transformation in them.

When the parade had marched off, I went to talk to the band. Battalion bands had long gone. Reduced numbers of bandsmen had been merged to form large bands and the duty band that day came from the Queen's Division (a grouping of infantry regiments of which the Royal Anglians were part). After speaking to the Director of Music, I turned to see the Pickwickean figure of 'Big John'

Millgate. When I commanded the Pompadours, there had been four Millgate brothers in the band and their father had been a former bandmaster of the battalion. 'Big John' was now a warrant officer and while he was large, he was also the Combined Services hockey goalkeeper. He was a strong, fit man. I walked over to him anticipating the usual salute. Instead, he reached forward and gave me a rib-crushing hug!

'You frightened the shit out of me, sir. How are you?'

I was amazed at just how much he cared and was even more surprised to learn how closely my progress and setbacks had been monitored by former Pompadours. It was my turn to control emotion.

At some time in the year I met Lieutenant General (retired) Sir James Wilson at an Army football match. He had a lifetime's involvement with a youth charity, formerly known as 'The Boy's Clubs' before it became 'Clubs for Young People' – a name change made with the admission of girls and young women. Sir James told me that the organisation needed a chief executive and that he thought I was admirably suited and qualified for the post. I applied and after being shortlisted and surviving two interviews at its London headquarters, I got the job. I would be the National Director of a UK-wide, youth organisation, with a membership of some 250,000 young people, in about 3,000 clubs, from 1 April 1996.

I handed over as CITG at Christmas. After that I was due resettlement and gardening leave until I left the service on 30 April. Sue had previously told me that she anticipated that I would get a job in London and that she wanted to live in West Byfleet – where we had lived during my MOD tour – because she had retained friends and connections there. She argued that she had followed the 'drum' for twenty-seven years and now it was her turn to choose. She had a point.

House hunting in Surrey from Wiltshire was never going to work, so we decided to rent a property for six months close to West Byfleet that would make our search easier. We moved to Ottershaw at the end of March and soon found and bought a wing of a Victorian convent in Woking, which was being redeveloped. It was set in three acres of attractive grounds. The convent was so large that the developer was able to divide it into twelve houses and ten apartments. He deservedly won the UK's best conversion award for 1996. Oldfield Wood was to prove a great place to live.

My final day in uniform came in the last week of March. I spent most of it with the Poachers, who were based in Warminster, with David Clements, my first adjutant, now their commanding officer. I was still a Deputy Colonel of the Regiment and could justify my visit. David took me to see some live firing on Salisbury Plain. 3 Platoon were carrying out platoon attacks and 3 Platoon had been my first platoon in Ballykinler in 1963. On a whim, I asked if I could join in and was given a rifle, a helmet and some webbing. On my last active day of service I became a rifleman, crawling across the ground, firing reasonably aimed shots and screaming at the top of my lungs as the platoon charged the enemy. The young soldiers must have thought I was completely mad. No matter – it was my last 'hoorah'. After drinks in the sergeants' mess and lunch with the officers, I was driven out of camp and out of regimental soldiering in the turret of a *Warrior*.

A day later I handed in my kit to the stores at Upavon. It was a huge moment for me, but to the elderly civilian storeman it was a common occurrence. My neatly folded field uniforms and carefully cleaned equipment were tossed onto various piles behind his counter. And that was that, but not quite. 'ID card please, sir.' I took the card out of my wallet. I had carried one for thirty-four years and it represented who I was. I pushed it across the counter's top and then it and the military part of me was gone.

Although I was acutely aware that it was coming, the abruptness and downbeat end of my army career had not been foreseen. I was not looking for trumpets and fanfares, but the ordinariness of a dingy store, presided over by a bored storeman, was somehow too bleak a finale for a third of a century's service. I felt incomplete and, for a while, I drifted in a void. While I was determined not to dwell on the past, neither could I ignore it. I allowed myself a short time to reflect on the Army and on my career.

There were remarkably few downsides to my service, but I had long been aware of three over-arching failings. First, the Army had been too slow to adapt and align more closely with changing values in civilian life, which had resulted in an ever-widening gap. Secondly, it's slowness to recognise the value of women soldiers had had professional repercussions and resulted in the Army's service women being kept separate in their own corps until the 1990s, when they were finally integrated with mainstream regiments and corps. And finally, however inadvertent it might have been, its treatment of military wives as second class citizens was shocking.

I accept that military values cannot align precisely with those held in civilian life because of the extremes soldiers have to confront. However, more could have been done to make the service more transparent and understandable. That would have required much better communication than my service had seen. During the campaign against the IRA and in the Falklands War, the Army actually moved in the opposite direction. While an IRA threat persisted, soldiers were not allowed to wear uniform outside barracks (security was put forward as the reason) and apart from newsreel footage, the Army became invisible to the civilian population in everyday life. That state of affairs was compounded when news bulletins in the Falklands War were given by a senior civil servant, who had the delivery of a 'speak your weight' machine. Why an authoritative military figure did not give the news confused people and lost some of the public support gained when the task force sailed. Thankfully, the Army has moved on from there – witness brigadiers standing alongside the Prime Minister to deliver part of recent Covid-19 briefings.

A consequence of continued poor communication was that the Army was undervalued. The meagre pay, equally inadequate equipment and poor accommodation were testimony to that – failings that only began to be addressed in the late 1970s and have made stuttering progress ever since. Sombre remembrance on Armistice Day and warm words when the Army is committed to rescue people from flooding, or for rolling out of testing and vaccination stations, are not enough. It would seem to many knowledgeable observers that post the Cold War, cuts and the taking of savings have been too deep. While new areas have been recognised (e.g. cyber warfare) and there is an ambitious programme to replace many aging, major weapons systems, size still matters. The chilling point made to me by a Soviet general in 1990 about the Army being too small (see page 279) is arguably even more apposite today than it was three decades ago, especially if it is to fulfil the global strategic role that the government has set for it.

Notwithstanding all of that, the lows were easily outweighed by the many incredible highs. At the top of my list are mutual trust; a determination and ability to get things done; the immediate comradeship amongst soldiers; and the many wonderful, enduring friendships that I had forged. I miss the training, the confidence it brings and the esprit de corps. However, the Army taught me how to generate those things and I took them with me to civilian life. I left recognising the interest and reward of the fascinating places that my service had taken me to – over forty countries visited – and the challenge and excitement that the frequent changes of direction and role that my career had seen. All of them hold memories that will never leave me and nearly a quarter of a century on, burn as brightly as ever.

Now, for the first time in my adult life I no longer had a rank, or the authority that went with it. I did not even have a number that had been so much part of me and which had identified me as a soldier. 'Mr Groves' sounded insipid, but it was who I was now.

My soldiering had been exciting, interesting, at times frustrating but tremendously fulfilling. I was entirely satisfied that my service had been thoroughly worthwhile. I had had a ball.

Life in uniform had been anything but boring. Now, what's next Mr Groves?

ACKNOWLEDGEMENTS

WRITING A FIRST book was a daunting task and I would not have been successful in completing it without the help of many wonderfully kind people and professional organisations. I would like to acknowledge and thank them here. They are:

Julia Wheeler – *Times* and BBC correspondent, who I met when she was part of the Cheltenham Book Festival aboard the *Queen Mary* in late 2019. Her encouragement and particularly her long list of helpful advice, set the ball in motion and then kept it rolling.

The Arvon Foundation – The first of Julia's advice was to attend a residential writing course organised by the Arvon Foundation. It was excellent.

Mel McGrath and Jay Griffiths – Highly accomplished authors and tutors on the Arvon course. They honed what literary skills I already possessed and gave me new tools and an understanding of how to take on the task.

The Royal Anglian Regimental Headquarters and Museum – for ready help and the excellence of and easy access to its online archives.

Chris Wilson – fellow second tenor in the Weybridge Male Voice Choir. He was an acquaintance at the beginning (now a friend) with no real knowledge of me or the Army. He read every early draft to help me gauge what interest the book might hold and how understandable it might be for a civilian audience.

Rory McClean and Marnie Summerfield Smith – Once the book was drafted Rory (another acclaimed author) helped me improve the draft. Subsequently, the indefatigable Marnie (she describes herself as a memoire mentor) was able to place the book with a publisher.

Ryan Gearing and the Unicorn Publishing Group – Together they made the publishing process easy for me.

Claire Davies – Fellow student on the Arvon course, who voluntarily became my editor. Meeting her was an enormous stroke of luck. Her enthusiasm and detailed, constructive criticisms added so much to the quality of the text and took her hours of concentrated work. We have become lifelong friends.

My Family – The memoire is as much about my immediate family as it is about me. My sons, Nick and Tim, have developed as the men I most admire and trust. That speaks volumes as I am fortunate enough to have many wonderful friends. Their support and confidence in me, during the writing of the book, never wavered. Sue simply continued to be my constant companion, loving wife and soulmate. It is impossible to put a value on her part in this story and in my life in general. I am a lucky, lucky, lucky man.

To all of you – I offer my grateful thanks.

GLOSSARY

ADC – Aide De Camp
AFA – Army Football Association
APC – Armoured Personnel Carrier
APTC – Army Physical Training Corps
ASD – Army Staff Duties
ATO – Ammunition Technical Officer
ATR – Army Training Regiment
BAOR – British Army of the Rhine
BBGT – Brigade and Battlegroup Trainer
BMH – British Military Hospital
BOAC – British Overseas Airways
  Corporation
CDS – Chief of the Defence Staff
CGS – Chief of the General Staff
CITG – Commander Initial Training
  Group
CO – Commanding Officer
CPX – Command Post Exercise
CSM – Company Sergeant Major
DDR – Deutshe Demokratische Republik
DemSM – Demonstrations Sergeant
  Major
FAC – Forward Air Controller
FIBUA – Fighting in Built Up Areas
FLOSY – Federation for the Liberation
  of the South Yemen
FFR – Fitness For Role
FTX – Formation Training Exercise
GCE – General Certificate of Education
GCSE – General Certificate of Secondary
  Education
GSO – General Staff Officer
HF – High frequency
ID – Identity (card)
IED – Improvised Explosive Device

IO – Intelligence Officer
IRA – Irish Republican Army
JDSC – Junior Division of the Staff
  College
JNCO – Junior Non Commissioned
  Officer
KD – Khaki Drill (clothing)
LAD – Light Aid Detachment
LSGC – Long Service and Good Conduct
  (Medal)
MACA – Military Assistance to Civil
  Authorities
MACC – Military Aid to the Civil
  Community
MACM – Military Aid to Other [Civil]
  Ministries/Departments
MBA – Master of Business
  Administration
MCC – Marylebone Cricket Club
MFO – Military Forwarding
  Organisation
MOD – Ministry of Defence
NAAFI – Naval, Army, Air Force
  Institute
NATO – North Atlantic Treaty
  Organisation
NBC – Nuclear, Biological, Chemical
NCO – Non Commissioned Officer
NITAT – Northern Ireland Training and
  Advisory Team
NLF – National Liberation Front
NORTHAG – Northern Army Group
NRA – National Rifle Association
OP – Observation Post
OPEC – Organisation of the Petroleum

Exporting Countries

PMC – President of the Mess Committee

POW – Prisoner Of War

RCB – Regular Commissions Board

R&R – Rest and Recuperation (leave)

RAF – Royal Air Force

RAAF – Royal Australian Air Force

RAEC – Royal Army Education Corps

RAMC – Royal Army Medical Corps

REME – Royal Electrical and Mechanical
Engineers

RMAS – Royal Military Academy
Sandhurst

RMCS – Royal Military College of
Science

RMP – Royal Military Police

RSM – Regimental Sergeant Major

RUC – Royal Ulster Constabulary

SASC – Small Arms School Corps

SLR – Self Loading Rifle

SNCO – Senior Non Commissioned
Officer

SPG – Special Patrol Group

STC – Sennelager Training Centre

STUFT – Ships Taken Up From Trade

TEWT – Tactical Exercise Without
Troops

UKLF – United Kingdom Land Forces

UN – United Nations

VHF – Very High Frequency

WRAC – Women's Royal Army